Emergent Gauge Symmetries in Particle Physics and Cosmology

EMERGENT GAUGE SYMMETRIES IN PARTICLE PHYSICS AND COSMOLOGY

Steven D. Bass

Kitzbühel Centre for Physics, Austria
Jagiellonian University in Krakow, Poland

World Scientific

NEW JERSEY · LONDON · SINGAPORE · BEIJING · SHANGHAI · TAIPEI · CHENNAI

Published by

World Scientific Publishing Co. Pte. Ltd.

5 Toh Tuck Link, Singapore 596224

USA office: 27 Warren Street, Suite 401-402, Hackensack, NJ 07601

UK office: 57 Shelton Street, Covent Garden, London WC2H 9HE

Library of Congress Control Number: 2025931739

British Library Cataloguing-in-Publication Data
A catalogue record for this book is available from the British Library.

ISBN 978-981-98-1147-2 (hardcover)
ISBN 978-981-98-1148-9 (ebook for institutions)
ISBN 978-981-98-1149-6 (ebook for individuals)

For any available supplementary material, please visit
https://www.worldscientific.com/worldscibooks/10.1142/14265#t=suppl

Desk Editor: Carmen Teo Bin Jie

Typeset by Stallion Press
Email: enquiries@stallionpress.com

To Magdalena, Céline and Cornelia

Preface

Particle physics today is characterised by the tremendous success of the Standard Model including the discovery of the Higgs boson, the only known (seemingly) elementary scalar particle. With gravitation there are exciting developments with new tests of General Relativity which so far works very well in all our observations. Highlights include the recent discovery of gravitational waves and the imaging of black hole horizons. Mysteries remain however, especially with explaining the tiny neutrino masses, the matter-antimatter asymmetry in the Universe, fermion families in the Standard Model, dark energy, dark matter, and also the physics associated with primordial inflation.

This book concerns the origin of the Standard Model and the gauge symmetries that govern its interactions. The Standard Model with its measured parameters is mathematically consistent up to the Planck scale where quantum gravity ideas might apply. At the same time, we do need some extra new ingredients to resolve the open physics puzzles waiting to be explained. Interestingly, if we extrapolate the Standard Model up towards energies approaching the Planck scale then we find that the stability of the Higgs vacuum is very sensitive to the exact values of Standard Model's masses and couplings. Perhaps there are important correlations between the Standard Model's parameters determined by new dynamics deep in the ultraviolet.

A possible solution involves an emergent Standard Model. Emergent gauge symmetries are well known in quantum condensed matter physics. Why not also in particle physics? The idea here is that the Standard Model including its particles and interactions might be born in some topological like phase transition at a scale about 10^{16} GeV. The gauge symmetries and particles of the Standard Model would then "dissolve" in the extreme ultraviolet. The Standard Model particles would be the stable collective long-range excitations of more primordial degrees of freedom that operate above the scale of emergence. With preference for a Minkowski vacuum, the cosmological constant then comes out nicely with an energy scale similar in size to the tiny masses of light Majorana neutrinos. There are also interesting possibilities for dark matter and baryogenesis in this scenario. Possible dark matter candidates include light mass axions, primordial black holes and relic fluctuations of the system above the scale of emergence that may have been frozen out in an initial phase transition in the early Universe that produced the Standard Model. Time dependent dark energy can be accommodated. This book serves as an introduction to this physics and as an invitation to the reader to further explore the theory and possible phenomenology of an emergent particle physics Standard Model, this plus its consequences for cosmology.

Next generation experiments will soon yield fresh new data pushing the frontiers in our quest to understand the deep structure of matter. These include the high luminosity upgrade of the LHC, new precision experiments plus cosmology surveys. Further, gravitational waves measurements and probes of polarisation in the Cosmic Microwave Background will enable us to look for signals from the early Universe. New developments with quantum sensors will play an essential role with measuring tiny effects with small signals in the experiments. These are exciting times. If the Standard Model is emergent, then there are deep connections between the infrared and ultraviolet waiting to be revealed.

For background reading, excellent textbooks on quantum field theory are [Aitchison and Hey (2013a); Bjorken and Drell (1965); Itzykson and Zuber (1980); Peskin and Schroeder (1995); Pokorski (2000); Taylor (1979); Weinberg (1995, 1996)]. The Standard Model and its phenomenology are well covered in [Aitchison and Hey (2013b); Altarelli (2013a)] with Cosmology surveyed in [Weinberg (2008)]. Recent reviews on the Higgs boson and its implications are given in [Bass *et al.* (2021); Jakobs and Zanderighi (2023)]. Excellent texts on General Relativity include [Dirac (1996); Einstein (1956); Misner *et al.* (1973); Straumann (2013); Weinberg (1972)].

I thank Markus Ackermann, Albert De Roeck, Michael Doser, Georgi Dvali, Klaus Helbing, Karl Jakobs, Fred Jegerlehner, Marumi Kado, Janina Krzysiak, Pawel Moskal, Stefan Pokorski, Sebastian Sapeta, Grigory Volovik and Erez Zohar for useful discussions on aspects of the physics topics discussed in this book. I thank my commissioning editor Annalisa Fischer and physics editors Soh Yong Qi and Carmen Teo Bin Jie at World Scientific Publishing Company for support and patience with the manuscript.

Contents

Chapter 1

Introduction

Our present understanding of particle physics and gravitation is provided by the Standard Model and General Relativity. The particle physics Standard Model works very well so far in describing the results of experiments from high energy physics at the Large Hadron Collider (LHC) to low energy precision measurements including the fine structure constant and the electron electric dipole moment. Its interactions underpin atomic, condensed matter and nuclear physics. The Standard Model is built on the local gauge symmetries of electroweak interactions and Quantum Chromodynamics (QCD), with the gauge groups $U(1)_Y$, $SU(2)_L$ and $SU(3)_c$. The spin-$\frac{1}{2}$ leptons and quarks interact through the exchange of spin-one gauge bosons. These are the massless photon of Quantum Electrodynamics, QED, the massless gluons of QCD and the massive W and Z bosons that mediate the weak interactions. Particle masses are generated by the Brout–Englert–Higgs mechanism. The Higgs boson with spin zero and mass 125 GeV discovered at CERN behaves very Standard Model like and completes the particle spectrum of the Standard Model. It is the first (and so far only) discovered (seemingly) elementary particle with spin zero [Bass *et al.* (2021); Jakobs and Zanderighi (2023)].

General Relativity describes gravitation wherever it has been tested. The recent discovery of gravitational waves [Abbott *et al.* (2016b)] and imaging of supermassive black hole horizons [Akiyama *et al.* (2019a, 2022)] are pushing the frontiers of gravity science. In the laboratory Newton's law has been shown to work down to at least 52 μm [Lee *et al.* (2020)]. General Relativity also behaves as a gauge theory with the physics invariant under local transformations of the coordinate system [Kibble (1961); Sciama (1964)]. General Relativity is a purely classical theory, unlike the Standard Model which is a relativistic quantum field theory.

The Standard Model and General Relativity have passed the tests of present observations. So far there is no evidence for new particles or interactions in the range of present experiments. Yet we know this is not the complete story. Some extra new physics is needed to understand issues with neutrinos (their masses, whether they might be their own antiparticles and possible CP phases), baryogenesis (why there is more matter than antimatter), the absence of strong CP violation linked to topological properties of gluon fields, fermion families (why are there three?), dark energy (the accelerating expansion of the Universe) and dark matter as well as primordial inflation.[1] The Standard Model particles that we study in laboratory experiments contribute just 5% of the energy budget of the Universe measured in cosmology with the remaining 68% residing in dark energy and 27% residing in dark matter.

How high in energy might the Standard Model work before one runs into new physics? The energy scale of any new interactions is so far not known. Where do the gauge symmetries of the Standard Model come from at a deeper level? What about the dynamics behind electroweak symmetry breaking? How should we understand the scale hierarchies of particle physics? The cosmological constant scale 0.002 eV is very much less than the electroweak scale 246 GeV, which is itself very much less than the Planck scale $M_{\mathrm{Pl}} = \sqrt{\hbar c / G} = 1.2 \times 10^{19}$ GeV where quantum gravity effects might apply. Quantum effects naively act to push the cosmological constant scale and the Higgs mass m_h towards very large values in contrast to the values measured in experiments. (Here in the definition of M_{Pl}, $\hbar$, c and G are Planck's constant, the speed of light and Newton's constant respectively.) The large Planck scale compared to Standard Model particle physics scales also feeds into the strengths of the different interactions. The dimensionless gravitational analogue of the QED fine structure constant $\alpha = \frac{1}{4\pi\epsilon_0} \frac{e^2}{\hbar c} \approx \frac{1}{137}$ is $\alpha_G = \frac{G M_{\mathrm{P}}^2}{\hbar c} = \frac{M_{\mathrm{P}}^2}{M_{\mathrm{Pl}}^2} \approx 5.9 \times 10^{-39}$ with M_{P} the proton mass 938 MeV. Gravitational effects are negligible in the domain of laboratory based particle physics experiments.

---

[1] In particle physics one also has few standard deviation anomalies awaiting explanation, "discrepancies" between measurements and Standard Model predictions that are not (yet) at the five standard deviations "discovery level" for new physics. Key examples are in flavour physics [Križan (2023)] and with the muon $g-2$ value [Jegerlehner (2017)]. The muon $g - 2$ anomaly enters when comparing to predictions based on dispersion relation theory. There is much better agreement with the Standard Model when compared to QCD lattice calculations [Boccaletti *et al.* (2024); Borsanyi *et al.* (2021)].

Today a vigorous programme of experiments plus theory is probing the high energy and precision frontiers. Together with cosmology surveys, the experiments are looking for cracks in our description of Nature provided by the Standard Model and General Relativity. Ideas for new physics and related experiments are discussed in contributions to the volume [Bass *et al.* (2023)] and the reviews in the Particle Data Group collection [Navas *et al.* (2024)]. Astroparticle investigations are covered in [Ackermann and Helbing (2023)]. In addition, next generation gravitational waves experiments [Bailes *et al.* (2021)] and measurements of the Cosmic Microwave Background polarisation [Dunkley (2015); Kamionkowski and Kovetz (2016); Komatsu (2022)] will be sensitive to possible new physics that might have been active in the early Universe. These experiments will provide an exciting new tool in our quest to understand the deep structure of matter.

While so far not revealing evidence for new particles and interactions, LHC data do suggest an intriguing result concerning the stability of the Standard Model vacuum. If we assume no coupling to new particles at higher energies and extrapolate the Standard Model into the deep ultraviolet using renormalisation group evolution, then the vacuum remains stable up to very high scales with a positive Higgs boson self-coupling — up to energy scales at least 10^{10} GeV and perhaps up to the Planck scale. This vacuum stability is very sensitive to the exact values of the parameters of the Standard Model [Bednyakov *et al.* (2015); Jegerlehner (2014c)]. Suppose we accept the high energy extrapolation. If one takes vacuum stability as an important principle, then one finds intriguing correlations between the Standard Model's parameters, its masses and couplings, which become connected to physics deep in the ultraviolet.

Might the Standard Model be more special than previously assumed? Here, in this book, we consider this possibility. In particular, we discuss the idea that the particles and interactions of the Standard Model might be emergent below a scale characterising some topological like phase transition deep in the ultraviolet [Bass (2021, 2023); Bjorken (2001a); Jegerlehner (2014c, 2019, 2021)]. In this scenario, the Higgs mass is environmentally selected, connected to the stability of the vacuum, with a subtle connection between the infrared world of our experiments and physics at work in the extreme ultraviolet. Further and interestingly, the cosmological constant scale comes out similar to the size of tiny Majorana neutrino masses [Bass and Krzysiak (2020a, 2020b)]. Majorana neutinos means that the neutrinos are their own antiparticles. New global symmetry breaking interactions will occur in higher dimensional operators, contributions which are suppressed by powers of the large scale of emergence. We will argue that with emergence this scale should be taken about 10^{16} GeV. The higher dimensional operators may have been active in the very early Universe when particle energies approached this large value. Their resulting effect might be relevant to cosmology observations today. If the Standard Model saturates the physics at mass dimension four up to the scale of emergence (meaning that any new interactions should be suppressed by this large scale), then dark matter might consist either of black holes formed in the early Universe or be associated with new physics (perhaps new particles) coupled through higher dimensional operators with just small non-gravitational coupling to normal Standard Model matter.

What do we mean by emergence? Emergence in physics occurs when a many-body system exhibits collective behaviour in the infrared that is qualitatively different from that of its more primordial constituents as probed in the ultraviolet [Anderson (1972); Palacios (2022)]. For example, classical physics is emergent from quantum physics. Hadrons like protons, neutrons and pions are emergent from quark-gluon degrees of freedom. Chemistry and biology are emergent from electrodynamics. An interesting case of emergent phenomena from everyday experience is the collective change in the travel direction of starling flocks from individual bird's flight fluctuations. Symmetries can also be emergent. As an everyday example of emergent symmetry, consider a carpet which looks flat and translational invariant when looked at from a distance. Up close, e.g. as perceived by an ant crawling on it, the carpet has structure and this translational invariance is lost. The symmetry perceived in the infrared, e.g. by someone looking at it from a distance, "dissolves" in the ultraviolet when the carpet is observed close up.

When seeking to understand the origin of symmetries it is important to distinguish between local gauge symmetries and the global symmetries of the theory. Local gauge symmetries act on internal degrees of freedom and not on the physical Hilbert space which is gauge invariant. In contrast, global symmetries relate eigenstates of the physical Hamiltonian. That is, local gauge symmetries are properties of the description of a system and global symmetries are properties of the system itself. Physical observables are invariant under the gauge symmetries of the Standard Model.

Emergent gauge symmetries are seen in quantum condensed matter systems. Gauged quasiparticles can arise with extra gauge symmetries beyond the photon exchanges associated with fundamental QED interactions between atoms and electrons. Emergent gauge symmetries are connected with quantum topological phase transitions. These phase transitions occur without a local order parameter and are instead associated with topological order as well as with long range quantum entanglement [Affleck *et al.* (1988); Baskaran and Anderson (1988); Levin and Wen (2005a); Moessner and Moore (2021); Powell (2020); Sachdev (2016); Volovik (2003, 2008); Wen (2004); Zaanen and Beekman (2012)]. Topological phase transitions differ from usual Landau–Ginzburg type phase transitions that do come with local order parameters. Collective gauge fields can "emerge" from the quantum structure of the many-body ground state. The ground state can exhibit multiple degeneracy with the degenerate substates being related by emergent gauge transformations.

In condensed matter physics the prototype for emergent gauge symmetries is the Fermi–Hubbard model of strongly correlated electrons in a two dimensional atomic lattice at half filling. This system exhibits spin-charge separation with the SU(2) spin of the electrons becoming dynamical with an emergent gauge symmetry beyond the more fundamental QED [Affleck *et al.* (1988); Baskaran and Anderson (1988)]. A further example is the A-phase of superfluid ^{3}He [Volovik (2003, 2008)]. This phase exhibits Fermi points — singular points in momentum space where the fermion quasiparticles have no mass gap. Close to the Fermi points the quasiparticles of ^{3}He-A are chiral fermions plus emergent spin-one gauge bosons and spin-two effective "gravitons". One finds emergent local gauge interactions with spin becoming dynamical to internal observers. The gauge symmetries correspond to the freedom (degeneracy) in choosing the position of the Fermi point on the former Fermi surface. The quasiparticles here each come with a common limiting velocity like what happens with Lorentz invariance in the Standard Model. There is also an emergent metric plus an analogue of the chiral anomaly. A recent overview of analogies between phenomena in gravitation and condensed matter physics is given in [Volovik (2023)]. Emergent gauge symmetries also play an important role in high temperature superconductors [Sachdev (2016)], the quantum Hall effect [Tong (2016)] as well as spin ice phenomena in magnetic systems [Rehn and Moessner (2016)]. Their application to quantum simulations of quantum field theories is discussed in [Bañuls *et al.* (2020); Banerjee *et al.* (2012); Zohar *et al.* (2016)].

How might the particle physics Standard Model be emergent?

Consider a statistical system near its critical point. The long range asymptote is a renormalisable quantum field theory with properties described by the renormalisation group [Peskin and Schroeder (1995); Wilson and Kogut (1974)]. If the spectrum includes $J = 1$ excitations among the degrees of freedom in the low energy phase, then it is a gauge theory. (Renormalisable quantum field theories with vector particles exhibit local gauge symmetries ['t Hooft (1980b)].) Gauge symmetries would then be an emergent property of the low energy phase and "dissolve" in some topological like phase transition deep in the ultraviolet [Bass (2021, 2023); Bjorken (2001a); Forster *et al.* (1980); Jegerlehner (1978, 1998, 2014c); 't Hooft (2007)]. The quarks and leptons as well as the gauge bosons and Higgs boson would be the stable collective long-range excitations of some (unknown) more primordial degrees of freedom that exist above the scale of emergence. The vacuum of the low energy phase should be stable below the scale of emergence.

Suppose the Standard Model works like this. Then it behaves as an effective theory with the renormalisable theory at mass dimension four, $D = 4$, supplemented by a tower of non-renormalisable higher dimensional operators each suppressed by powers of the large scale of emergence. The global symmetries of the Standard Model at $D = 4$ are constrained by gauge invariance and renormalisability. The higher dimensional operators are less constrained and may exhibit extra global symmetry breaking. Lepton number violation and tiny Majorana neutrino masses may enter at $D = 5$ [Weinberg (1979)], meaning that they are suppressed by a single power of the scale of emergence. One finds neutrino masses $m_\nu \sim \Lambda_{\mathrm{ew}}^2/M$ with Λ_{ew} the electroweak scale and M the scale of emergence. Possible proton decays might occur at $D = 6$ suppressed by two powers of M [Weinberg (1979); Wilczek and Zee (1979)]. New CP violation, needed for baryogenesis, might occur in Majorana neutrino phases at $D = 5$ as well as in new $D = 6$ operators [Grzadkowski *et al.* (2010)].

Within this emergence picture if one increases the energy much above the electroweak scale, then the physics becomes increasingly symmetric with energies $E \gg \Lambda_{\mathrm{ew}}$ until we come within about 0.1% or so of the scale of emergence. Then new global symmetry violations in the higher dimensional operators become important so the physics becomes increasingly chaotic until one goes through the phase transition associated with the scale of emergence, with the physics above this scale then described by new degrees of freedom and perhaps new physical laws. This scenario contrasts with unification models which involve maximum symmetry in the extreme ultraviolet. In unification models the gauge couplings of the Standard Model would meet at some large scale. If one takes the minimal Standard Model with no extra interactions, then the gauge couplings almost meet but not quite. LHC data (so far) reveal no evidence for higher dimensional correlations in searches for new operator terms divided by powers of a large mass scale below the few TeV range [Ellis *et al.* (2021); Slade (2019)]. The tiny neutrino masses indicated by neutrino oscillation experiments suggest a scale of emergence deep in the ultraviolet.

In addition to topological-like phase transitions, there are also ideas where emergent gauge symmetries can appear through the decoupling of gauge violating terms in the renormalisation group at an infrared fixed point (where couplings become invariant under changes in the scale or resolution) [Wetterich (2017)] and also in connection with possible spontaneous breaking of Lorentz symmetry, SBLS [Bjorken (2001a, 1963, 2010); Chkareuli *et al.* (2001)]. In the former case, the coefficient of any local gauge symmetry violating terms blows up at the fixed point, in contrast to restoration of global symmetries where the coefficient of any symmetry violating term goes to zero at the fixed point. With SBLS, non observability of any Lorentz violating terms at $D = 4$ corresponds to gauge symmetries for vector fields like the photon. Possible Lorentz violation might be manifest in terms at largest of $\mathcal{O}(\Lambda_{\text{ew}}^2/M^2)$ with a preferred reference frame naturally identified with the frame where the Cosmic Microwave Background is locally at rest [Bjorken (2001a)].

When coupling to gravitation emergence gives a simple explanation of the cosmological constant and dark energy. One also finds interesting constraints on possible dark matter scenarios. The cosmological constant Λ counts the energy density of the vacuum perceived by gravitation. Within General Relativity it provides the simplest explanation of dark energy. It is connected with the symmetries of the metric $g_{\mu\nu}$. With a finite cosmological constant Einstein's equations of gravitation have no vacuum solution where $g_{\mu\nu}$ is the constant Minkowski metric. That is, global spacetime translational invariance of the vacuum is broken by a finite value of Λ [Weinberg (1989)]. The reason is that a non-zero cosmological constant acts as a gravitational source which generates a dynamical spacetime with accelerating expansion of the Universe for positive Λ. Suppose that the vacuum including condensates with finite vacuum expectation values is spacetime translational invariant and that flat spacetime is consistent at mass dimension four, just as suggested by the success of the Standard Model. With the Standard Model as an effective theory emerging in the infrared, the low-energy global symmetries including spacetime translation invariance can be broken through additional higher dimensional terms suppressed by powers of the large scale of emergence M. QCD and electroweak interactions are characterised by the scales $\Lambda_{QCD} \approx 300$ MeV and $\Lambda_{ew} \approx 246$ GeV. These scales might then enter the cosmological constant with the scale of the leading term suppressed by the factor Λ_{ew}/M — see [Bass and Krzysiak (2020a, 2020b)] and the early work [Bjorken (2001a, 2001b)]. This scenario, if manifest in nature, would explain why the cosmological constant scale $\mu_{vac} = 0.002$ eV is similar to what we expect for neutrino masses based on neutrino oscillation measurements [Altarelli (2005)]. In the emergence picture the neutrinos are expected to be Majorana particles. One finds $\mu_{vac} \sim m_\nu \sim \Lambda_{ew}^2/M$. The precision of global symmetries in our experiments, e.g. lepton and baryon number conservation, tells us that in this scenario the scale of emergence should be deep in the ultraviolet, much above the Higgs boson and other Standard Model particle masses. Taking the value $\mu_{vac} = 0.002$ eV from astrophysics together with $\Lambda_{ew} = 246$ GeV gives a value for M about 10^{16} GeV. The tiny cosmological constant then involves a cancellation between zero-point energies, potential terms in the vacuum and a "bare gravitational" term consistent with the symmetries of the metric. This "bare gravitational term" can be thought of as parametrising the effect of physics above the scale of emergence. An interesting analogy in condensed matter physics is the Gibbs–Duhem relation for quantum liquids. Here, the zero-point energy from low temperature quasiparticles is cancelled by the effect of macroscopic degrees of freedom above the characteristic energy for the quantum liquid effective theory [Volovik (2005)].

The scale 10^{16} GeV is within the range where the Higgs self-coupling λ might cross zero (if indeed it does) if the Standard Model is extrapolated up to the highest energy scales using renormalisation group evolution. It is also similar to the "GUT scale" that typically appears in unification models. If the Standard Model is emergent at a scale where λ is non-negative then its vacuum will be fully stable. Any perturbative extrapolation of Standard Model degrees of freedom above the scale of emergence would reach into an unphysical region. For an emergent Standard Model, an interesting question involves the critical dimension for any phase transition that produces it. Might $3+1$ dimensions be special? Interestingly, the scale 10^{16} GeV is also typically taken as the scale of inflation in models of the early Universe. Perhaps emergence and inflation might be connected (?) If dark energy is time dependent, this might correspond to time dependence in the scale of ultraviolet completion for the Standard Model and/or in the coefficient of the cosmological constant term that appears in the low energy expansion in terms of higher dimensional operators. This time dependence would reflect the relaxation of the Standard Model away from the early Universe phase transition that produced it.

If the Standard Model saturates particle physics phenomena at $D = 4$ up to the scale of emergence, possible new physics including new particle couplings might be residing in higher dimensional operators. Dark matter candidates would then include light mass axion particles as well as long range fluctuations of the system above the scale of emergence that might have frozen out together with emergent Standard Model gauge interactions in the early Universe. A condensed matter analogy for the latter might come from extension of the Fermi–Hubbard model. In its low energy limit at half filling with spin-charge separation the physics involves emergent $SU(2)$ and $U(1)$ gauge boson quasiparticles coupled to the electron degrees of freedom. Any extra collective vibrations of the atomic lattice sites would behave as long range bosonic fluctuations — phonons — with a possible analogy for dark matter. Primordial black holes might also be an important source of dark matter. New sources of CP violation can occur in higher dimensional operators. This CP violation plus departure from thermal equilibrium at the emergence phase transition might provide the source needed for baryogenesis in the early Universe.

Given the well known challenges with quantising gravity, should it really be quantised at the Planck scale or might General Relativity be emergent at a scale below M_{Pl}? If the gauge symmetries of particle physics might be emergent from more primordial physics, it seems not implausible that those of gravitation might be also.

This book explores present thinking on these topics: possible emergent gauge symmetries in particle physics and the resulting consequences for cosmology. The plan of the book is as follows. Chapter 2 focuses on the very successful Standard Model built on the gauge theories of QED, QCD and electroweak interactions. The connections between gauge and Lorentz invariance are discussed here, as is the Brout–Englert–Higgs mechanism for generating gauge boson masses and charged fermion masses in the Standard Model. Following this introductory material, Chapter 3 discusses the issue of vacuum stability with energy scale dependent running couplings. Here one finds possible correlations between Standard Model parameters and the physics of the deep ultraviolet. Chapter 4 concerns quantum field theory anomalies and the connection between the physics measured in experiments and the symmetries of the extreme ultraviolet that enter through regularisation and renormalisation. Chapters 5 and 6 explain the concept of emergent gauge symmetries. Chapter 5 focuses on the idea of an emergent particle physics Standard Model. Here, we discuss the accompanying tower of higher dimensional operators with important physics being Majorana neutrino masses as well as possible proton decays and axion particles plus possible extra CP violation beyond the minimal Standard Model. In Chapter 6 we discuss quantum condensed matter systems where emergent gauge symmetries are observed in connection with topological phase transitions and long range quantum entanglement. Chapters 7 and 8 are about gravitation and cosmology. General Relativity is our present theory of gravitation. There is need for extra dark energy and dark matter to understand cosmology and astrophysics observations. In Chapter 9 we explain how emergence ideas offer a simple interpretation of the cosmological constant puzzle: Why is the cosmological constant so small? Emergence ideas can help with understanding the scale hierarchies observed in particle physics: the Higgs mass hierarchy puzzle and the size of the cosmological constant. These are discussed in Chapter 10. Chapter 11 continues this discussion with how emergence ideas might help explaining open puzzles in particle physics including dark matter and baryogenesis. Lastly, in Chapter 12, we bring together the key ideas and the challenges for experiment and theory with looking for signals of possible emergent gauge symmetries in particle physics and cosmology.

Chapter 2

The Standard Model:
Gauge Symmetries in Particle Physics

The Standard Model is built from Quantum Electrodynamics, QED, plus weak interactions and Quantum Chromodynamics, QCD, with the local gauge symmetries of $U(1)_Y$, $SU(2)_L$ and $SU(3)_c$. These ingredients enter together with the global symmetries of Poincaré invariance — that is, invariance under translations, rotations and Lorentz boosts. The Standard Model provides an excellent description of particle physics processes measured in current experiments from low energy precision measurements up to high energy colliders. The gauge symmetries determine the dynamics and interactions between elementary particles. Gauge interactions are mediated via the exchange of spin-one bosons: photons, W and Z weak bosons and gluons.

The different interactions come with very different phenomenology. QED is in the Coulomb phase with massless photons. QCD is confining with the phenomena of asymptotic freedom connecting quarks and massless gluons in the ultraviolet and colour singlet hadrons in the infrared. Weak interactions are in the Higgs phase with massive W and Z gauge bosons plus an accompanying scalar Higgs particle. The Higgs boson discovered at CERN in 2012 behaves very Standard Model like and completes the spectrum of the Standard Model. With the measured Higgs boson mass, the theory has good consistent ultraviolet behaviour: it is renormalisable and satisfies perturbative unitarity with a vaccum that is (close to) stable under quantum corrections without the need for extra interactions. The weak interactions maximally violate parity with the $SU(2)_L$ gauge symmetry acting only on left-handed fermions.

Atoms are built from electrons (the lightest charged lepton) and a nucleus composed of protons and neutrons, each built of light up and down quarks, all fermions with spin $\frac{1}{2}$. The gauge bosons which mediate the Standard Model interactions carry spin one. With confinement the range of QCD gluon exchange interactions is about 1 fm (10^{-15}m). At longer distances quarks and gluons are always confined inside of hadrons. Weak interactions operate on a distance scale about 0.01 fm with the W and Z bosons having masses of 80 and 91 GeV. The weak interactions power the Sun and nuclear reactors and in the process radiate neutrinos into the final state.

The fermions of the Standard Model come in three families (or generations) with heavier quarks and leptons in addition to the light up and down quarks plus the electron and its partner neutrino. In all, the minimal Standard Model (before tiny neutrino masses) has 18 parameters: 3 gauge couplings and 15 in the Higgs sector (6 quark masses, 3 charged lepton masses, 4 quark mixing angles including one CP violating complex phase, the W and Higgs boson masses). Our everyday experience is described by just the first generation of light quarks and leptons (protons, neutrons, pions, ... built from up and down quarks, electrons, plus neutrinos from the Sun). However, our existence is not insensitive to the physics at the highest energy scales. The stability of the Higgs vacuum after quantum corrections needs the top quark with its large mass ≈ 173 GeV when we take into account vacuum polarisation and associated renormalisation group effects with energy scale dependent running couplings.

In this chapter, we describe the different parts of the Standard Model. Its high energy behaviour and vacuum stability issues are discussed in Chapter 3.

2.1   Quantum Electrodynamics

Quantum Electrodynamics, QED, follows from the gauge principle of requiring the physics to be invariant under local U(1) transformations of the phases of the fields describing electrically charged particles, e.g. the electron and proton.

QED with electrons and photons is described through the Lagrangian

$$\mathcal{L} = \bar{\psi}\big(i\gamma^{\mu}D_{\mu} - m\big)\psi - \frac{1}{4}F_{\mu\nu}F^{\mu\nu}. \tag{2.1}$$

Here, ψ represents the electron field and $D_\mu \psi = (\partial_\mu + ieA_\mu)\psi$ is the gauge covariant derivative with A_μ the photon field and e the electric charge. The fine structure constant is $\alpha = e^2/4\pi$ and $F_{\mu\nu} = \partial_\mu A_\nu - \partial_\nu A_\mu$ is the electromagnetic field tensor. The photon is massless and the electron comes with mass 0.51 MeV.[1]

QED dynamics follow from requiring invariance under the local U(1) gauge transformation $\psi \to e^{i\omega(x)}\psi$ where $\omega(x)$ is a function of the spacetime coordinates. Derivative terms ∂_μ acting on ψ will also act on the phase factor $\omega(x)$ so that this phase factor does not just flow through the combination $\partial_\mu \psi$. Instead, one considers the gauge covariant derivative

$$\partial_\mu \mapsto D_\mu = \partial_\mu + ieA_\mu \qquad (2.2)$$

with the photon field transforming as $A_\mu \to A_\mu - \frac{1}{e}\partial_\mu\omega(x)$ so that $D_\mu\psi \to e^{i\omega(x)}D_\mu\psi$. The Lagrangian is then invariant under the combined local transformations of ψ and A_μ. Maxwell's equations are derived from the photon's equations of motion

$$\partial_\mu F^{\mu\nu} = j^\nu \qquad (2.3)$$

with $j^\nu = ie\bar{\psi}\gamma^\nu\psi$. These include Gauss's Law $\nabla \cdot \mathbf{E} = \rho$ where $\mathbf{E}$ is the electric field $\mathbf{E} = -\partial\mathbf{A}/\partial t - \nabla A_0$ and $\rho = ie\psi^\dagger\psi$. The corresponding magnetic field is $\mathbf{B} = \nabla \times \mathbf{A}$ with both $\mathbf{E}$ and $\mathbf{B}$ gauge transformation invariant.

The Feynman rules for perturbative QED and related Feynman diagrams follow from quantisation of the electron and photon fields in Eq. (2.1), see e.g. [Bjorken and Drell (1965)].

Gauge invariance is interconnected with Lorentz symmetry. In general, A_μ does not transform as a four-vector under Lorentz transformations but is supplemented by an additional gauge term. Let $U(\epsilon)$ denote an infinitesimal unitary Lorentz transformation. One finds

$$U(\epsilon)A_\mu(x)U^{-1}(\epsilon) = A_\mu(x') - \epsilon_{\mu\nu}A^\nu(x') + \frac{\partial\Lambda(x',\epsilon)}{\partial x'^\mu}, \qquad (2.4)$$

where Λ is an operator gauge function. Gauge invariance ensures that the action remains Lorentz invariant. The structure of Eq. (2.4) ensures that gauge invariant Maxwell equations are Lorentz covariant [Bjorken and Drell (1965); Weinberg (1995)].

---

[1]Present experimental constraints put the photon's mass $m_\gamma < 1 \times 10^{-18}$eV and electric charge $q_\gamma < 1 \times 10^{-35}e$ [Navas *et al.* (2024)].

Real photons come with two transverse polarisations whereas A_μ has also time and longitudinal components which are really not dynamical degrees of freedom. For example, canonical quantisation singles out A_0 at the expense of manifest covariance with the fields defined at equal time. The time component A_0 commutes with all operators and is a c-number and not an operator, in contrast with the space components A_i. Gauss's law then implies that $\nabla \cdot \mathbf{A}$ is also a c-number [Bjorken and Drell (1965)]. One makes some selection of "gauge fixing" (or constraint on the gauge fields) which defines the dynamical degrees of freedom in the book keeping of the calculation. By a suitable choice of gauge — the radiation or Coulomb gauge $\nabla \cdot \mathbf{A} = 0$ and $A_0 = 0$ — only transverse degrees of freedom of the photon field appear in calculations though at the expense of losing manifest Lorentz and gauge covariance in the formalism.[2] Observables such as S-matrix elements are Lorentz covariant and independent of the gauge fixing procedure.

One can also use a manifestly covariant gauge fixing procedure, generalising the Lorentz gauge choice from classical electrodynamics $\partial_\mu A^\mu = 0$. Here, the Lagrangian is supplemented with the gauge fixing term $\delta\mathcal{L} = -(\partial_\mu A^\mu)^2/2\xi$ where $1/\xi$ acts as a Lagrange multiplier field. These gauge choices are known as the family of R_ξ gauges. The limit $\xi \to 0$ is known as Landau gauge. The Feynman–'t Hooft gauge corresponds to $\xi = 1$. S-matrix elements are independent of ξ. Within R_ξ gauges one still has to deal with longitudinal degrees of freedom. This is solved by the Gupta–Bleuler procedure where the gauge fixing is imposed via the matrix element constraint $\partial^\mu A_\mu^+ |\Psi\rangle = 0$ for physical states $|\Psi\rangle$ where A_μ^+ is the (annihilation) positive frequency part of the photon field $A_\mu(x) = \int d^3k \sum_\lambda \{\epsilon_\mu^\lambda(\mathbf{k})a(\mathbf{k},\lambda)e^{-ik.x} + (\epsilon_\mu^\lambda(\mathbf{k}))^*a^\dagger(\mathbf{k},\lambda)e^{ik.x}\}$ or, equivalently, by taking $\langle\Psi| \, \partial^\mu A_\mu \, |\Psi\rangle = 0$. Here $a(\mathbf{k},\lambda)$ and $a^\dagger(\mathbf{k},\lambda)$ are the photon annihilation and creation operators for photons with momentum $\mathbf{k}$ and polarisation λ, and ϵ_μ^λ are the corresponding polarisation vectors. The vacuum state $|\text{vac}\rangle$ is defined with $a(\mathbf{k},\lambda)|\text{vac}\rangle = 0$ corresponding to no net photons.

---

[2]With canonical quantisation in $A_0 = 0$ gauge Gauss's Law becomes an operator constraint rather than a classical equation of motion. The spatial components of the vector potential and components of the electric field are canonical pairs. However, the time component A_0 does not have a canonical pair. The quantisation procedure treats it differently as a Lagrange multiplier that imposes Gauss's Law as an operator constraint. This guarantees that the physical Hilbert space contains only physical degrees of freedom.

2.1.1 *Global and discrete symmetries*

Relativistic quantum field theories include the global symmetries of the Poincaré group: translations, rotations and Lorentz boosts. Spin-statistics properties with commutation relations for bosons and anticommutation relations for fermions in $3 + 1$ dimensions follow for a theory with Lorentz covariance and a unique ground state if the condition of microscopic causality is to be satisfied [Bjorken and Drell (1965); Feynman (1999)].

The QED Lagrangian is invariant under global U(1) and (for massless electrons) chiral U(1) transformations. Noether's theorem tells us that there are conserved currents associated with each of the continuous global symmetries of the classical Lagrangian. Translational invariance is associated with momentum conservation. Rotational invariance is associated with angular momentum conservation. Electric charge conservation is linked to global U(1) invariance in QED with the vector current as the corresponding Noether current. The electron mass term breaks the chiral symmetry between left- and right-handed electrons which are symmetric in their interactions with photons. If we neglect the electron mass, then the left- and right-handed electron fields $\psi_L = \frac{1}{2}(1 - \gamma_5)\psi$ and $\psi_R = \frac{1}{2}(1 + \gamma_5)\psi$ transform independently under chiral rotations. Helicity is conserved for massless electrons. Invariance under global chiral rotations of the phase of the fermion fields $\psi \rightarrow e^{i\omega\gamma_5}\psi$ leads to the Noether current $J_{\mu 5} = \bar{\psi}\gamma_\mu\gamma_5\psi$. Conservation of $J_{\mu 5}$, which also corresponds to electron helicity conservation, is softly broken by the electron mass term.[3]

In addition to global symmetries, the discrete symmetries of the theory also constrain the dynamics. Discrete symmetries are built from combinations of charge conjugation invariance, C, parity transformations, P, and time reversal, T. The QED interactions described by the Lagrangian Eq. (2.1) respect the discrete symmetries C, P, T and the combinations CP and CPT. The combination CPT is a fundamental property of local quantum field theories with a Hermitian Hamiltonian, invariance under proper Lorentz transformations and spin-statistics [Bjorken and Drell (1965); Sozzi (2008)]. CPT invariance holds independent of possible violations of individual discrete symmetries, e.g. like P and CP violation found with weak interactions.

---

[3]It is also sensitive to anomalous terms in its divergence equation in the singlet channel where one couples through two gauge boson intermediate states — a quantum effect associated with a clash of classical symmetries in Feynman loop diagrams [Adler (1969); Bell and Jackiw (1969)] (see Chapter 4).

2.1.2   *Running couplings*

When vacuum polarisation associated with quantum fluctuations is taken into account, the coupling $\alpha = \frac{e^2}{4\pi}$ depends on the resolution or momentum transfer in the interaction. This scale dependence is determined by the renormalisation group equation for the "running" scale dependent $\alpha(\mu^2)$

$$\mu^2 \frac{d}{d\mu^2} \frac{\alpha}{4\pi} = \beta_{\text{QED}}(\alpha). \tag{2.5}$$

Here, μ is the renormalisation scale that enters in any calculation where we have to control the divergences associated with quantum corrections (Feynman loop diagrams) (see Chapter 4). This scale can be set equal to the momentum scale that we probe with. S-matrix elements depend on external momenta but are independent of the renormalisation scale (or how a theoretician set up the calculation) whereas the individual values of α and the electron mass m_e appearing in the calculation do depend on the renormalisation scale when we go beyond simple tree approximation, viz. beyond the simple Born term. The perturbative expansion for the so-called β−function β_{QED} reads

$$\beta_{\text{QED}}(\alpha) = n_f \frac{4}{3} \left(\frac{\alpha}{4\pi} \right)^2 + 4n_f \left(\frac{\alpha}{4\pi} \right)^3 + \cdots, \tag{2.6}$$

where n_f is the number of species of fermions that participate in the fermion loop diagrams. Because of the positive sign for $\beta_{\text{QED}}(\alpha)$ the scale dependent coupling $\alpha(\mu^2)$ increases logarithmically with increasing scale μ^2.

The fine structure constant is defined as

$$\alpha(m_e^2) = \frac{1}{4\pi\epsilon_0} \frac{e^2}{\hbar c} \approx \frac{1}{137} \tag{2.7}$$

with ϵ_0 the electric permitivity of the vacuum, i.e. through probing the electron with a photon carrying zero energy and with zero momentum transfer. For $n_f = 1$, the running α at scales λ_{UV} above m_e is related to $\alpha(m_e^2)$ in leading order approximation through [Landau *et al.* (1956); Landau and Pomeranchuk (1955)]

$$\alpha(\lambda_{\text{UV}}^2) = \frac{\alpha(m^2)}{1 - \frac{\alpha(m^2)}{3\pi} \ln \frac{\lambda_{\text{UV}}^2}{m^2}}. \tag{2.8}$$

This expression has a singularity called the Landau pole in the mathematical limit that $\lambda_{\text{UV}} = m \, e^{3\pi/2\alpha(m^2)}$, which has a value very much larger than the Planck mass $M_{\mathrm{Pl}} = 1.2 \times 10^{19}$ GeV. Numerically, the Landau pole for QED with just electrons, $n_f = 1$, and photons evaluates to be $\lambda_{\mathrm{UV}} = 5 \times 10^{276}$ GeV at leading-order or 10^{227} GeV if one extends the calculation to next-to-leading-order. In the full Standard Model, summing over charged lepton and quark contributions reduces the Landau pole scale to 10^{34} GeV [Gockeler *et al.* (1998)].

It remains an open question whether the theory has an ultraviolet fixed point where $\alpha(\mu^2)$ might freeze at some finite value deep in the ultraviolet due to non-perturbative dynamics when α becomes large. For discussion see [Weinberg (1995)] and [Pokorski (2000)]. Key issues include dynamical chiral symmetry breaking at large values of the coupling α [Gockeler *et al.* (1998); Kogut *et al.* (1988)] and possible extra non-leading terms, suppressed by powers of the large ultraviolet scale of completion if QED is treated as an effective theory [Djukanovic *et al.* (2018)].

2.1.3 *Precision QED*

Quantum Electrodynamics is our most accurately tested theory, presently with precision of one part in 10^{12}. A key observable is the fine structure constant. One compares the value of $\alpha(m_e^2)$ extracted from measurements of the electron's anomalous magnetic moment $a_e = (g - 2)/2$ (and combined with QED theory) with measurements from atom interferometry experiments with Caesium, Cs, and Rubidium, Rb. In QED, the electron's a_e is given by a perturbative expansion in α which is known to $\mathcal{O}(\alpha^5)$ precision plus tiny QCD and weak interaction corrections. The present most accurate measurement of a_e comes with precision 0.13×10^{-12} [Fan *et al.* (2023)], viz.

$$a_e^{\mathrm{exp}} = 0.001\ 159\ 652\ 180\ 59\ (13). \tag{2.9}$$

QED perturbation theory using Feynman diagrams involving just electrons and photons gives [Aoyama *et al.* (2018)]

$$
\begin{aligned}
a_e^{\mathrm{QED,e}} = {} & \frac{\alpha}{2\pi} - 0.328\ 478\ 965\ 579\ 193 \cdots \left(\frac{\alpha}{\pi}\right)^2 \\
& + 1.181\ 241\ 456\ 587 \cdots \left(\frac{\alpha}{\pi}\right)^3 \\
& - 1.912\ 245\ 764 \cdots \left(\frac{\alpha}{\pi}\right)^4 + 6.675\ (192) \left(\frac{\alpha}{\pi}\right)^5 + \cdots, \tag{2.10}
\end{aligned}
$$

where we write $\alpha \equiv \alpha(m_e^2)$.[4] There are also extra loop contributions with heavy leptons

$$a_e(\text{QED} : \mu, \tau) = 2.747\ 571\ 9(13) \times 10^{-12} \tag{2.11}$$

as well as electroweak and QCD corrections

$$a_e^{\text{SM}} = a_e^{\text{QED}} + 0.030\ 53\ (23) \times 10^{-12}\ (\text{weak})$$
$$+ 1.692\ 7\ (120) \times 10^{-12}\ (\text{hadronic}) \tag{2.12}$$

which need to be added to get the full Standard Model prediction. The present experimental accuracy is sensitive to the heavy lepton and QCD radiative corrections but not yet to weak interaction effects. The electron a_e value then determines a precision measurement of α (modulo any radiative corrections from new physics beyond the Standard Model, BSM). Taking the theoretical formulae Eqs. (2.10)–(2.12) enables one to extract α from a_e^{SM}. One finds

$$1/\alpha|_{a_e^{\text{SM}}} = 137.035\ 999\ 166\ (15). \tag{2.13}$$

Atom interferometry experiments measure heavy Cs or Rb atomic masses through recoil of a Cs or Rb atom in an atomic lattice. The fine structure constant $\alpha^2 = (4\pi R_\infty/c)(m_{\text{atom}}/m_e)(\hbar/m_{\text{atom}})$ then involves this atomic mass measurement combined with other experimental quantities: the Rydberg constant R_∞ and the ratio of the atom to electron masses m_{atom}/m_e; c is the speed of light and $\hbar$ is Planck's constant). Any radiative corrections from BSM physics will enter a_e but not the atomic physics measurements of α. Thus, comparing the different determinations of α gives a precision test of QED as well as constraining possible new physics scenarios.

Substituting the measurements of α extracted from Cs [Parker *et al.* (2018)]

$$1/\alpha|_{\text{Cs}} = 137.035\ 999\ 046\ (27) \tag{2.14}$$

and Rb [Morel *et al.* (2020)]

$$1/\alpha|_{\text{Rb}} = 137.035\ 999\ 206\ (11) \tag{2.15}$$

tabletop atom interferometric experiments into the QED expansion for a_e gives agreement to one part in 10^{12}.

---

[4]A small uncertainty has been reported in the coefficient of the $\mathcal{O}(\alpha^5)$ term [Volkov (2019)]. The effect of this second calculation would be just to change the final digits "66" to "59" in Eq. (2.13).

CPT symmetry predicts the same mass for the electron and positron.[5] It also says that the magnetic moment should change sign between electrons and positrons. The successful agreement of the electron a_e and atomic physics measurements of α as a precision test of QED to at least one part in 10^{12} points to the implicit working of CPT at this level with CPT a fundamental property of QED. More explicitly, the C and CPT symmetries have been shown to work to the level of 2×10^{-12} in measurements of $g-2$ for both electrons and positrons with [Van Dyck *et al.* (1987)]

$$g(e^-)/g(e^+) = 1 + (0.5 \pm 2.1) \times 10^{-12}. \qquad (2.16)$$

2.2 Non-abelian Quantum Chromodynamics

The gauge principle arguments connecting dynamics with invariance under local rotations of the phase that led to QED can be readily generalised to non-abelian gauge groups, e.g. SU(2) and SU(3) — so called Yang–Mills theories.

In Quantum Chromodynamics (QCD), the fermions form an SU(3) triplet labelled by the colour label red, green and blue, with the physics invariant under local rotations in SU(3) colour space

$$\psi \to \mathcal{G}\psi, \qquad (2.17)$$

where

$$\mathcal{G}(x) = e^{i\frac{1}{2}\vec{\lambda}\cdot\vec{\omega}(x)} \qquad (2.18)$$

is an element of the gauge group and describes rotating the local colour phase of the quark fields. In Eq. (2.18), λ_a are the eight 3×3 Gell-Mann matrices with $\frac{1}{2}\lambda_a$ the generators of SU(3). The QCD gauge covariant derivative is

$$D_\mu \psi = \left[\partial_\mu + ig_s \frac{\lambda_a}{2} A_\mu^a\right]\psi, \qquad (2.19)$$

where g_s is the SU(3) colour charge and A_μ^a are the gluon fields. Under local SU(3) gauge transformations the gauge covariant derivative transforms as

$$D_\mu \to \mathcal{G}D_\mu\mathcal{G}^{-1} \qquad (2.20)$$

---

[5]Taking the values quoted by the Particle Data Group [Navas *et al.* (2024)], the difference in the electron and positron masses is presently constrained as $(m_{e^+} - m_{e^-})/m_{\text{average}} < 8 \times 10^{-9}$ at 90% confidence level with the sum of their electric charges $|q_{e^+} + q_{e^-}|/e < 4 \times 10^{-8}$.

with $A_\mu = \frac{1}{2}\vec{\lambda} \cdot \vec{A}_\mu$ transforming as

$$A_\mu \to A'_\mu = \mathcal{G} A_\mu \mathcal{G}^{-1} + \frac{i}{g_s}(\partial_\mu \mathcal{G})\mathcal{G}^{-1}. \qquad (2.21)$$

The gluon field tensor

$$G^a_{\mu\nu} = [D_\mu, D_\nu]_-/(ig_s) = \partial_\mu A^a_\nu - \partial_\nu A^a_\mu - g_s f_{abc} A^b_\mu A^c_\nu \qquad (2.22)$$

then induces non-abelian 3 and 4 gluon interaction vertices with the gluons also carrying colour charge and not just the quarks. Here, the f_{abc} are the structure constants of SU(3), $[t^a, t^b] = if_{abc}t^c$ with $t^a = \frac{1}{2}\lambda^a$. Note the fundamental difference to QED where, unlike the electron, the photon has no electric charge.

Putting things together, the QCD Lagrangian for a quark with mass m is

$$\mathcal{L}_{QCD} = \bar{\Psi} i\gamma^\mu D_\mu \Psi - m\bar{\Psi}\Psi - \frac{1}{4}\mathrm{Tr}\ G_{\mu\nu}G^{\mu\nu}, \qquad (2.23)$$

where one takes the trace over gluon colour indices. The Lagrangian $\mathcal{L}_{QCD}$ is invariant under the SU(3) gauge transformations in Eqs. (2.17) and (2.21). If we write the quark field ψ as the sum of left- and right-handed quark components $\psi = \psi_L + \psi_R$, where $\psi_L = \frac{1}{2}(1 - \gamma_5)\psi$ and $\psi_R = \frac{1}{2}(1 + \gamma_5)\psi$ project out different states of quark helicity, then the QCD Lagrangian becomes

$$\mathcal{L}_{QCD} = \bar{\Psi}_L i\gamma^\mu D_\mu \Psi_L + \bar{\Psi}_R i\gamma^\mu D_\mu \Psi_R - m(\bar{\Psi}_L \Psi_R + \bar{\Psi}_R \Psi_L)$$

$$- \frac{1}{4}\mathrm{Tr}\ G_{\mu\nu}G^{\mu\nu}. \qquad (2.24)$$

For massless quarks the left-handed and right-handed quark fields transform independently under unitary global chiral rotations — so called chiral symmetry. In the QCD Lagrangian just the scalar mass term connects left- and right-handed quarks.

Quantisation and Feynman diagrams are obtained using the techniques either of canonical, path integral or stochastic quantisation. With canonical quantisation the fields are treated as operators satisfying canonical commutation relations. Green's functions are calculated as vacuum expectation values of products of the field operators with anti-commutation relations taken for the creation/annihilation operators for fermions and commutation relations for bosons [Bjorken and Drell (1965)]. With Feynman path integral quantisation the fields are c-numbers and the Lagrangian is taken as classical. The Green's functions follow from integrating the product of fields over the sum of all their possible functional forms, that is over all Feynman paths, with a suitable weight, see e.g. [Itzykson and Zuber (1980); Muta (1987); Peskin and Schroeder (1995); Pokorski (2000); Taylor (1979); Weinberg (1995)]. Stochastic quantisation relies on similarity between functional integral expressions for Green's functions in Euclidean space and statistical averaging with the fields taken as stochastic variables [Parisi and Wu (1981)]. QCD perturbation theory and its applications are reviewed in [Huston *et al.* (2024); Muta (1987)].

With Yang–Mills quantisation extra Fadeev–Popov ghost fields are introduced to preserve consistency of the theory and to maintain unitarity. These ghost fields violate usual spin statistics (they come with spin zero and anti-commute) and arise only in loop diagrams as virtual particles with no free states.[6]

QCD comes with a rich phenomenology. The coupling α_s satisfies asymptotic freedom whereby it decreases logarithmically with increasing resolution. One also finds non-trivial gauge topology. Only hadrons, colourless bound states of quarks and gluons, exist as physical states in the ground state spectrum.

2.2.1 *Asymptotic freedom*

With gluons carrying colour charge, the 3 gluon interaction vertex induces asymptotic freedom. The QCD coupling $\alpha_s(Q^2) = g_s^2/4\pi$ — the QCD version of the fine structure constant for quark-gluon and gluon-gluon interactions — decreases logarithmically with increasing resolution or increasing four-momentum transfer squared Q^2 that we probe the QCD system with [Gross and Wilczek (1973); Politzer (1973)]. This contrasts with QED where the running coupling $\alpha(Q^2)$ has a slow logarithmic rise with increasing Q^2.

At leading order

$$\alpha_s(Q^2) = \frac{g_s^2}{4\pi} = \frac{4\pi}{\beta_0 \ln(Q^2/\Lambda_{\text{QCD}}^2)}. \qquad (2.25)$$

Here, $\beta_0 = \frac{11}{3}N_c - \frac{2}{3}n_f$ where $N_c = 3$ is the number of colours and n_f is the number of active quark flavours; Λ_{QCD} is the renormalisation group invariant QCD infrared scale with $\Lambda_{\text{QCD}} \approx 300$ MeV for three light flavours.

---

[6]With the "wrong" spin statistics, they fail the conditions for causality and unitarity and so cannot create physical states.

The scale dependence of α_s is given by the renormalisation group equation

$$\mu^2 \frac{d}{d\mu^2} \frac{\alpha_s}{4\pi} = \beta_{\text{QCD}}(\alpha_s), \qquad (2.26)$$

where

$$\beta_{\text{QCD}}(\alpha_s) = -b_0 \left(\frac{\alpha_s}{4\pi}\right)^2 - b_1 \left(\frac{\alpha_s}{4\pi}\right)^3 - \cdots \qquad (2.27)$$

with

$$\begin{aligned}
b_0 &= \frac{11C_A - 4n_f T_R}{12\pi} = \frac{33 - 2n_f}{12\pi} \\
b_1 &= \frac{17C_A^2 - n_f T_R(10C_A + 6C_F)}{24\pi^2} = \frac{153 - 19n_f}{24\pi^2}.
\end{aligned} \qquad (2.28)$$

For QCD the group constants are $C_A \equiv N_c = 3$, $C_F = (N_c^2 - 1)/(2N_c) = \frac{4}{3}$ and $T_R = \frac{1}{2}$ with $t_{ab}^A t_{ab}^B = T_R \delta_{AB}$.

The key detail here is the different sign in the QCD β function contributions from quark fermion loops (terms proportional to n_f in Eq. (2.28)) and gluonic gauge boson loop terms. For QCD the gluon loops win and give us asymptotic freedom with a negative β_{QCD} function! The vanishing of $\beta_{\text{QCD}}(\alpha_s)$ with $\alpha_s \to 0$ when $Q^2 \to \infty$ is known as an ultraviolet renormalisation group fixed point. The QCD β–function $\beta_{\text{QCD}}(\alpha_s)$ has been evaluated to five-loop accuracy [Baikov *et al.* (2017)]. The value of α_s is usually quoted relative to its value at the Z boson mass scale, $\alpha_s(m_Z) = 0.1180(9)$ [Navas *et al.* (2024)]. The scale dependent running of α_s compared to the values extracted from high-energy collider experiments is shown in Fig. 2.1.

The perturbative expression for α_s has α_s rising without bound in the infrared, e.g. corresponding to the perturbative singularity at $Q^2 = \Lambda_{\text{QCD}}^2$ in Eq. (2.25). However, it is commonly believed that α_s freezes at some low infrared scale corresponding to an infrared fixed point $\beta_{\text{QCD}}(\alpha_s^{\text{IR}}) = 0$ due to non-perturbative confinement effects — see [Gribov (1987, 1999); Mandula (1977)] and related phenomenology in [Dokshitzer and Webber (1995); Mattingly and Stevenson (1994)].

2.2.2 *Gluon topology*

QCD Yang–Mills also comes with non-trivial gluon topology defined through non-local large-distance properties of gluon fields. Consider the

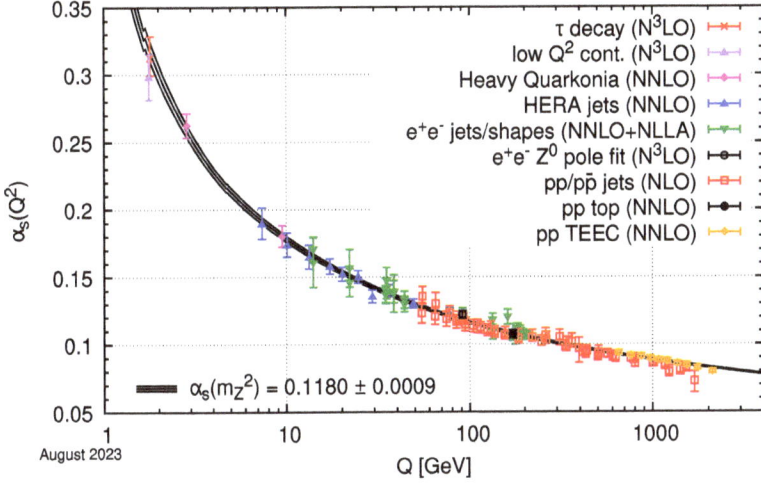

Fig. 2.1. Summary of determinations of α_s in the year 2023 as a function of the energy scale Q compared to the running of the coupling computed at five loops taking as input the current Particle Data Group average $\alpha_s(m_Z^2) = 0.1180 \pm 0.0009$. Figure from [Huston *et al.* (2024)].

the topological charge n defined through the equation

$$n = \int d^4x \, Q_t = \int d^4x \, \frac{\alpha_s}{8\pi} G_{\mu\nu} \tilde{G}^{\mu\nu}, \qquad (2.29)$$

where $G_{\mu\nu}$ is the gluon field tensor and $\tilde{G}^{\mu\nu} = \frac{1}{2} \epsilon^{\mu\nu\alpha\beta} G_{\alpha\beta}$. This quantity n is invariant under local deformations of the gluon field A_μ^a, viz.

$$\frac{\delta}{\delta A_\mu} n[A] = 0, \qquad (2.30)$$

and is therefore determined by non-local properties of A_μ^a. The gauge invariant topological charge density

$$Q_t = \frac{\alpha_s}{8\pi} G_{\mu\nu} \tilde{G}^{\mu\nu} \qquad (2.31)$$

is the total divergence of a gluonic current — the (gauge dependent) Chern–Simons current

$$K^\mu = \frac{g_s^2}{32\pi^2} \epsilon^{\mu\alpha\beta\gamma} A_\alpha^a \left(G_{\beta\gamma}^a - \frac{1}{3} g_s c^{abc} A_\beta^b A_\gamma^c \right). \qquad (2.32)$$

The topological charge is then associated with a finite surface integral

$$n = \int d^4x \, Q_t = \int d^4x \, \partial^\mu K_\mu = \oint_S d\sigma^\mu K_\mu, \qquad (2.33)$$

where the integral is taken over a hypersphere in four dimensions. The current K^μ is closely connected to the QCD axial anomaly discussed in Chapter 4. Finiteness of the classical action requires that

$$x^2 G^a_{\mu\nu}(x) \to 0 \text{ almost everywhere as } x_\alpha \to \infty. \tag{2.34}$$

This in turn implies that gluon fields satisfying the gauge transformation rule $\mathcal{G}$, $A_\mu \to \mathcal{G}A_\mu\mathcal{G}^{-1} + \frac{i}{g_s}(\partial_\mu\mathcal{G})\mathcal{G}^{-1}$, should tend to a pure gauge configuration

$$A_\mu \to \frac{i}{g_s}(\partial_\mu\mathcal{G})\mathcal{G}^{-1} \tag{2.35}$$

when $x_\alpha \to \infty$. One finds that the topological charge takes quantised values with either integer or (perhaps) fractional values — the topological winding number [Belavin *et al.* (1975); Crewther (1978)]. Gauge transformations come in two kinds: small gauge transformations that are topologically deformable to the identity (e.g. associated with perturbative QCD) and large gauge transformations associated with non-zero topological winding number [Shifman (1991)].

Unitary large gauge transformations change the winding number associated with any vacuum substate whereas the net vacuum should be invariant. Let $|m\rangle$ denote a vacuum substate characterised by topological winding number m. Under a large gauge transformation characterised by topological winding number n one finds

$$\mathcal{G}(g_{(n)})|m\rangle = |n+m\rangle. \tag{2.36}$$

Corresponding to this topology and beyond QCD perturbation theory, the net QCD vacuum is then a superposition of vacuum substates characterised by different topological winding numbers [Callan *et al.* (1976); Jackiw and Rebbi (1976)],

$$|\text{vac}, \theta\rangle = \sum_m e^{im\theta}|m\rangle \tag{2.37}$$

with

$$\mathcal{G}(g_{(n)})|\text{vac}, \theta\rangle = e^{-in\theta}|\text{vac}, \theta\rangle. \tag{2.38}$$

The substates $|m\rangle$ with individual topological winding number are defined with respect to a choice of gauge but the net vacuum $|\text{vac}, \theta\rangle$ is gauge invariant up to a phase [Pokorski (2000)]. The QCD vacuum is characterised by a value of θ which is a constant of motion with each possible θ value corresponding to the ground state of an independent sector of the Hilbert space. That is, for any given gauge invariant operator $\hat{O}$

$$\langle \text{vac}, \theta | \ \hat{O} \ | \text{vac}, \theta' \rangle = 0 \quad \text{if } \theta \neq \theta'. \tag{2.39}$$

When quarks are included the substates $|m\rangle$ also contain delocalised quark-antiquark pairs $q_L \bar{q}_R$. Each pair comes with finite chirality $\chi_q = \pm 2$, so that each contributing vacuum substate $|m\rangle$ has zero net axial charge (corresponding to the flavour-singlet axial vector current): $2mf - \sum_q \chi_q$ with f the number of active quark flavours (see Chapter 4). Instantons (tunneling processes that connect $|m\rangle$ states differing by winding number ± 1) correspond to hopping across the potential barrier between different vacuum substates enabling liberation (or absorbtion) of quark-antiquark pairs with non-zero chirality while conserving energy and momentum in quark scattering processes from instanton configurations [Crewther (1978); 't Hooft (1976a,b)].

It is not allowed to restrict the vacuum to just a specific substate $|m\rangle$ with invariance only under small but not large gauge transformations. Doing so would violate the principle of clusterisation, one of the fundamental properties of quantum field theory which is connected to causality and unitarity. Clusterisation means that the vacuum expectation value of the time ordered product of several local operators must be reducible to the sum over intermediate states including the vacuum intermediate state plus excitations over the given vacuum. Physically non-vanishing two point functions involving quark chirality operators would otherwise vanish without the Bloch superposition of substates in Eq. (2.37); for details see [Shifman (1991)].

The θ angle becomes a physical parameter only for massive quarks. Rotations of the vacuum angle θ and flavour-singlet chiral rotations are equivalent. This means that θ can be rotated to zero for the idealised case of exact chiral symmetry corresponding to massless quarks. In practice, chiral symmetry for light quarks is broken by small quark masses so it is not an exact symmetry. This means that we need to consider the possibility of finite θ values. If θ is finite, then it can induce new strong CP violation linked to QCD gluon topology effects (see Section 2.2.3). Experiments constrain $\theta < 10^{-10}$. Thus, in the real world, one has a genuine θ issue to solve. Understanding why θ is so small is one of the main open puzzles in particle physics. For an extended discussion of chiral properties of the θ-vacuum $|\text{vac}, \theta\rangle$ see [Crewther (1980)].

The QCD version of Gauss's Law is no longer linear in the gluon fields. From the gluon equation of motion

$$\partial^\mu G^a_{\mu\nu} = \bar{\psi}\gamma_\nu \lambda^a \psi - g_s f^{abc} G^{\mu\nu}_b A_{c\mu} \qquad (2.40)$$

setting $\nu = 0$ gives

$$\nabla \cdot \mathbf{E}^a = \rho^a_{\text{quark}} + g_s f^{abc} \mathbf{E}^b . \mathbf{A}^c \qquad (2.41)$$

with colour electric and magnetic fields $\mathbf{E}_i = G_{i0}$ and $\mathbf{B}_1 = G_{23}$ &tc. Gauss's Law in QCD is important in Hamiltonian lattice QCD which is formulated in the $A_0 = 0$ gauge [Creutz (1985)] with the Hamiltonian formulation also used in quantum simulations of quantum field theories [Bañuls *et al.* (2020)]. The net colour electric flux from each lattice site vanishes and ensures that the resulting Hilbert space is invariant under small (local) gauge transformations. The situation with large gauge transformations is discussed in [Grosse *et al.* (1997)] and [Halimeh and Hauke (2022)]. By itself, Gauss's Law requires physical states only to be invariant under small (but not under large) gauge transformations. Formally, small gauge transformations act as the generator of Gauss's Law.

Gauge fixing issues become more subtle with gluon topology. The gauge fixing condition $\partial^\mu A_\mu = 0$ does not uniquely specify the gauge fixing since one can have A_μ and A'_μ both satisfying the divergence equation and still be related by a large gauge transformation. One needs to remove these so-called Gribov copies. The resulting gluon propagator no longer has a Kallén–Lehmann spectral representation corresponding to a physical massless pole. The gluon is thus expelled from the spectrum of physical excitations [Gribov (1977, 1978)]; see also [Zwanziger (1989)] and the review [Vandersickel and Zwanziger (2012)].

2.2.3 *QCD Phenomenology: Confinement Physics*

QCD phenomenology works with quark and gluon degrees of freedom describing scattering processes with high momentum transfer giving way to colourless (colour singlet) hadrons as the external states in our experiments and the ground state degrees of freedom. Strong interactions confine the fundamental QCD degrees of freedom with large α_s coupling in the infrared. That is, one finds a change in degrees of freedom from coloured quarks and gluons to colour singlet hadrons as emergent excitations in the strongly coupled theory. These hadrons are baryons like the proton and neutron (fermions built of three valence quarks) and mesons (bosons built from a valence quark-antiquark pair) plus perhaps possible glueball states built from colour singlet combinations of gluons. For glueball states there are hints in data, though these states await definitive experimental confirmation.

QCD is our theory of strong interactions and the structure of hadrons. Historically, it developed from the Eightfold Way patterns observed in hadron spectroscopy with wavefunctions described in terms of spin $\frac{1}{2}$ quarks labeled by SU(3) flavour, SU(2) spin and, inside baryons, antisymmetric in a new SU(3) colour label to satisy the Pauli exclusion principle (with the wavefunctions symmetric in their flavour-spin labels) plus the parton description of deep inelastic scattering.[7] Then came the insight that colour is a dynamical quantum number and the discovery of QCD as a non-abelian local gauge theory with coloured gluons as the gauge bosons mediating interactions between quarks and gluons [Fritzsch and Gell-Mann (1972); Fritzsch *et al.* (1973)]. Asymptotic freedom [Gross and Wilczek (1973); Politzer (1973)] with the essential role of the non-abelian three gluon vertex provides the connection between high energy and low energy phenomenology with the QCD coupling $\alpha_s(Q^2)$ decreasing logarithmically with increasing momentum transfer Q^2, that is with small interaction strength in the ultraviolet and strong interactions in the infrared.

Low energy QCD is characterised by confinement and dynamical chiral symmetry breaking, DChSB. The long range gluon fields generate a confining potential that binds quarks always inside hadrons with a QCD confinement radius of order 1 fm $= 10^{-15}$ m.[8] The gluonic confinement potential contributes most of the proton's mass of 938 MeV with the rest determined by small quark mass perturbations. The masses of the proton's constituent two up quarks and one down quark are about 2.2 MeV for each up quark and 4.7 MeV for the down quark in the QCD Lagrangian. The colour hyperfine one-gluon-exchange potential (OGE), between confined quarks generates the mass splitting between the spin $\frac{1}{2}$ nucleon and its spin $\frac{3}{2}$ lowest mass

---

[7]The light SU(3) flavours are labeled up, down and strange (their flavour denoted u, d and s). These quarks carry electric charges $e_u = +\frac{2}{3}$ and $e_d = e_s = -\frac{1}{3}$ where, e.g. a proton is built from two up quarks and a down quark, and a neutron is built of two down quarks and an up quark. The spin-zero and spin-one mesons are built of a quark-antiquark combination. The quark flavours were later found also to include heavier charm, bottom and top quarks.

[8]Quarks are bound by a string of glue which can break into two colourless hadron objects involving the creation of a quark-antiquark pair corresponding to the newly created ends of two confining strings formed from the original single string of confining glue. There are no isolated quarks.

$\Delta(1232)$ resonance excitation [Close (1979)]. For massless quarks the QCD Lagrangian is chiral symmetric. When there is no quark mass term the left-handed and right-handed quark fields in the QCD Lagrangian transform independently under unitary chiral rotations

$$q_{\mathrm{L}} \to q_{\mathrm{L}}' = U_{\mathrm{L}} q_{\mathrm{L}}; \quad q_{\mathrm{R}} \to q_{\mathrm{R}}' = U_{\mathrm{R}} q_{\mathrm{R}}, \tag{2.42}$$

where here q_L and q_R denote SU(3) flavour multiplets of the quark fields. On the other hand, there is an absence of parity doublets in the light-hadron spectrum. For example, the $J^P = \frac{1}{2}^+$ proton and the lowest mass $J^P = \frac{1}{2}^-$ N*(1535) nucleon resonance are separated in mass by 597 MeV. This tells us that the chiral symmetry for light u and d (and s) quarks is dynamically (spontaneously) broken. DChSB is associated with formation of a scalar quark condensate in the vacuum connecting left- and right-handed quark fields, $\langle \mathrm{vac} \mid \bar{\psi}\psi \mid \mathrm{vac} \rangle < 0$. The lightest mass pseudoscalar mesons, the pion and kaon, are the would-be Nambu–Goldstone bosons [Goldstone (1961); Nambu (1960)] associated with DChSB and special with their mass squared proportional to the masses of their valence quarks inside, $m_P^2 \propto m_q$ [Gasser and Leutwyler (1982)].[9] Naively, one expects nine Nambu–Goldstone bosons associated with DChSB: eight with SU(3) and one with axial U(1) rotations. The isoscalar partners of the pion and kaon, the η and η' mesons, are too heavy to be pure Nambu–Goldstone states. These mesons are sensitive both to Nambu–Goldstone dynamics and to non-perturbative gluon topology active in the flavour singlet channel. This gluonic input supplies them with extra mass plus extra interaction through gluonic intermediate states.

The pion and kaon satisfy the Gell-Mann–Oakes–Renner (GMOR) relation [Gell-Mann *et al.* (1968)]

$$m_P^2 f_\pi^2 = -m_q \langle \mathrm{vac} \mid \bar{\psi}\psi \mid \mathrm{vac} \rangle + \mathcal{O}(m_q^2). \tag{2.43}$$

Here, the pion decay constant $f_\pi = \sqrt{2} F_\pi = 130$ MeV plays the role of the order parameter for DChSB. The lightest mass pions have mass 135 MeV for the π^0 and 140 MeV for the charged $\pi^\pm$. Substituting the pion masses and kaon masses ($m_{K^\pm} = 494$ MeV and $m_{K^0} = m_{\bar{K}^0} = 498$ MeV) then

---

[9]Goldstone's theorem tells us that there is one massless pseudoscalar boson for each symmetry generator that does not annihilate the vacuum, here the global symmetry transformations $\psi \to e^{i\gamma_5 \frac{\lambda_a}{2}\omega^a}\psi$ [Goldstone (1961); Goldstone *et al.* (1962)].

gives the leading-order quark mass ratios

$$\left.\frac{m_u}{m_d}\right|_{\mathrm{LO}} = 0.55, \quad \left.\frac{m_s}{m_d}\right|_{\mathrm{LO}} = 20. \tag{2.44}$$

The extra gluonic mass term in the flavour-singlet channel $\tilde{m}_{\eta_0}^2$ is associated with the Yang–Mills (pure glue theory) topological susceptibility [Veneziano (1979); Witten (1979)] with an essential role for non-trivial gluon topology [Shore (1998)]. The Witten–Veneziano formula $m_\eta^2 + m_{\eta'}^2 = 2m_K^2 + \tilde{m}_{\eta_0}^2$ relates $\tilde{m}_{\eta_0}^2$ and the η, η' and kaon masses at leading order in the chiral expansion. Phenomenologically, this expression works to $\approx 10\%$ with $\tilde{m}_{\eta_0}^2 = 0.73$ GeV^{-2} [Di Vecchia and Veneziano (1980)]. The Witten–Veneziano formula has also been confirmed in QCD lattice calculations to 10% accuracy [Cichy *et al.* (2015)]. The physics of gluonic contributions to the η' involves a subtle interplay of local anomalous Ward identities and non-local gluon topology. In general, η'-hadron interactions are observed to be characterised by large OZI violations enhanced by gluonic intermediate states [Bass and Moskal (2019)].

Hadron masses are connected to gluonic matrix elements via the trace anomaly in the QCD energy-momentum tensor [Collins *et al.* (1977); Shifman (1991)]. Whereas QCD with massless quarks is classically scale invariant, the proton mass is finite with infrared physics characterised by the infrared scale $\Lambda_{\mathrm{QCD}} \approx 300$ MeV associated with the running QCD coupling α_s, see Eq. (2.25). Scale/conformal transformations are associated with the scale or dilatation current $d^\mu = x_\nu \theta^{\mu\nu}$ with $\theta_{\mu\nu}$ the QCD energy-momentum tensor; d_μ satisfies the divergence equation $\partial_\mu d^\mu = \theta^\mu{}_\mu$ with

$$\theta^\mu{}_\mu = (1 + \gamma_m) \sum_q m_q \bar{\psi}_q \psi_q + \beta(\alpha_s)/4\alpha_s \; G^a_{\mu\nu} G^{\mu\nu}_a. \tag{2.45}$$

This is non-vanishing for massless quarks with $\beta(\alpha_s)/4\alpha_s \; G^a_{\mu\nu} G^{\mu\nu}_a$ the trace anomaly term. Here, γ_m is the quark mass anomalous dimension, $\mu^2 \frac{d}{d\mu^2} m_q = \gamma_m m_q$ with $\gamma_m = -\alpha_s/\pi + \cdots$, m_q is the renormalised quark mass and μ is the renormalisation scale; $\beta(\alpha_s)$ is the QCD β-function. The forward proton matrix element of $\theta_{\mu\nu}$ is $\langle p, s|\theta_{\mu\nu}|p, s\rangle = p_\mu p_\nu$ with trace $\langle p, s|\theta^\mu{}_\mu|p, s\rangle = M_{\mathrm{p}}^2$ relating the proton mass M_{p} squared to the gluonic trace anomaly term [Jaffe and Manohar (1990)]. (Here, p_μ denotes the proton's momentum vector and s_μ is its spin vector.) In contrast, pions and kaons would be massless in the chiral limit with massless quarks. Here the gluonic trace anomaly term must vanish. Internal binding cancels against individual quark-antiquark terms as manifest in e.g. the

Nambu–Jona-Lasino model [Klevansky (1992); Nambu and Jona-Lasinio (1961)] with the massless pions and kaons emerging as Nambu–Goldstone bosons. At low resolution the three valence quarks in the proton behave as massive constituent quark quasiparticles through interaction with the vacuum condensate produced by DChSB.

Besides $m_{\eta'}$, gluon topology also involves interesting physics linked to the size of the QCD θ angle. An important part of chiral rotations includes flavour-singlet axial U(1) rotations of the quark fields. With the θ vacuum containing delocalised quark-antiquark pairs, the vacuum state $|\mathrm{vac}, \theta\rangle$ transforms under flavour-singlet chiral rotations as [Pokorski (2000)]

$$e^{i\alpha\tilde{Q}_5}|\mathrm{vac}, \theta\rangle = |\mathrm{vac}, \theta + 2\alpha n_f\rangle, \qquad (2.46)$$

where the charge $\tilde{Q}_5$ is defined with respect to the partially conserved version of the flavour-singlet axial-vector current defined in Chapter 4 — see Eqs. (4.5) and (4.7). Invariance of the QCD θ-vacuum then requires that, in parallel to this singlet chiral rotation, θ should transform as $\theta \to \theta - 2\alpha n_f$ to compensate the angular rotation dependence. The effect of vacuum topological structure leads one to consider the most general QCD Lagrangian $\mathcal{L}_{\mathrm{QCD}} \to \mathcal{L}_{\mathrm{QCD}} + \theta_{\mathrm{QCD}}Q_t$ where the extra term $\theta_{\mathrm{QCD}}Q_t$ is odd under CP symmetry. If θ_{QCD} is non-zero this new Lagrangian term can induce strong interaction CP violation effects induced by topological properties of the gluon fields.

This physics is readily seen through the effective chiral Lagrangian for low energy QCD. Effective chiral Lagrangians are used to describe low energy QCD physics involving the would-be Nambu–Goldstone bosons plus the η'. These Lagrangians are constructed consistent with the chiral symmetries of the underlying QCD (and independent of the detailed gluon dynamics that generates the large η' mass term) [Di Vecchia and Veneziano (1980); Leutwyler (1998); Witten (1980)]. The role of the θ angle is manifest in the flavour-singlet sector. In the low-energy chiral Lagrangians the mesons are included via the unitary meson matrix $U = \exp i(\phi/F_\pi + \sqrt{\frac{2}{3}}\eta_0/F_0)$, where $\phi = \sum \pi_a \lambda_a$ denotes the octet of would-be Nambu–Goldstone bosons π_a, and η_0 is the singlet boson; F_0 is the singlet decay constant, which at leading order in the chiral expansion taken to be equal to $F_\pi = 92.2$ MeV. Under unitary chiral transformations, the meson matrix U transforms as $U \to U_R \times U \times U_L^\dagger$ corresponding to the chiral rotations U_L for left-handed quarks and U_R for right-handed quarks in Eq. (2.42). The gluonic mass contribution $\tilde{m}_{\eta_0}^2$ for the η' is introduced via a flavour-singlet potential involving the topological charge density Q_t

and is constructed so that the Lagrangian also reproduces the QCD axial anomaly (discussed in detail in Chapter 4). This potential reads

$$\frac{1}{2}iQ_t\mathrm{Tr}\left[\log U - \log U^\dagger\right] + \frac{3}{\tilde{m}_{\eta_0}^2 F_0^2}Q_t^2 \;\mapsto\; -\frac{1}{2}\tilde{m}_{\eta_0}^2\eta_0^2, \tag{2.47}$$

where Q_t is eliminated through its equation of motion to give the gluonic mass term $\tilde{m}_{\eta_0}^2$. Requiring invariance under global axial U(1) transformations $U \to e^{-2i\alpha}U$ suggests an extra term, $-\theta_{\mathrm{QCD}}Q_t$, in the effective Lagrangian for axial U(1) physics so that the potential in Eq. (2.47) becomes

$$\frac{1}{2}iQ_t\mathrm{Tr}\left[\log U - \log U^\dagger\right] + \frac{3}{\tilde{m}_{\eta_0}^2 F_0^2}Q_t^2 - \theta_{\mathrm{QCD}}Q_t \tag{2.48}$$

with the singlet chiral rotation being compensated by $\theta_{\mathrm{QCD}} \to \theta_{\mathrm{QCD}}-2\alpha n_f$.

Can the net value of θ_{QCD} in QCD be non-zero? For massless quarks θ_{QCD} can be rotated away by a suitable chiral rotation so that, in this case, it is not physical. (One just needs $m_u = 0$ for this to work, viz. rotating away any complex phase for the remaining mass terms plus rotating the θ angle to zero without introducing any extra complex mass term [Hook (2019)].) However, chiral dynamics tells us that the lightest up and down flavour quarks have small but finite masses [Gasser and Leutwyler (1982); Leutwyler (2013)]. Chiral symmetry breaking then means that one needs to consider possible finite θ values. When quark masses are included, the parameter that determines the size of strong CP violation is $\Theta_{\mathrm{QCD}} = \theta_{\mathrm{QCD}} + \mathrm{Arg}\det\mathcal{M}_q$, where $\mathcal{M}_q$ is the quark mass matrix.[10] That is, the most general QCD Lagrangian becomes

$$\mathcal{L}_{\mathrm{QCD}} \to \mathcal{L}_{\mathrm{QCD}} + \Theta_{\mathrm{QCD}}Q_t. \tag{2.50}$$

Possible strong CP violation then links QCD and the Higgs sector in the Standard Model, with the latter determining the quark masses.

If Θ_{QCD} has a non-zero value, then it induces a non zero neutron electric dipole moment $d_n = 5.2 \times 10^{-16}\,\Theta_{\mathrm{QCD}}$ ecm [Crewther *et al.* (1979)]. The neutron electric dipole moment d_n is a measure of possible CP violation

---

[10]The most general quark mass terms

$$\mathcal{L}_{\mathrm{mass}} = -\bar{q}_{\mathrm{L}i}\mathcal{M}_{ij}q_{\mathrm{R}j} - \bar{q}_{\mathrm{R}i}(\mathcal{M}^\dagger)_{ij}q_{\mathrm{L}j} \tag{2.49}$$

can be diagonalised via the chiral rotations in Eq. (2.43), viz. $q_{\mathrm{L}} \to q_{\mathrm{L}}' = U_{\mathrm{L}}q_{\mathrm{L}}$ and $q_{\mathrm{R}} \to q_{\mathrm{R}}' = U_{\mathrm{R}}q_{\mathrm{R}}$. Part of this transformation involves the $U_A(1)$ piece $q_{\mathrm{L}} \to q_{\mathrm{L}}' = e^{-i\alpha}q_{\mathrm{L}}$ and $q_{\mathrm{R}} \to q_{\mathrm{R}}' = e^{i\alpha}q_{\mathrm{R}}$ where $\alpha = \frac{1}{n_f}\mathrm{Arg}\det\mathcal{M}$ [Peccei (1999)].

from any new physics induced anisotropy in the vacuum polarisation of the neutron. The interaction with an electric field $\mathbf{E}$ is described by an interaction term $-\mathbf{d_E}.\mathbf{E}$ where $\mathbf{d_E}$ is the electric dipole moment. Under time reversal $\mathbf{E} \rightarrow \mathbf{E}$; $\mathbf{d_E}$ is proportional to the neutrons's spin vector which is odd under time reversal. Hence, any finite electric dipole moment corresponds to a violation of T symmetry and, through CPT, to a violation of CP [Sozzi (2008)]. Experiments constrain $|d_n| < 3.0 \times 10^{-26}$ecm at 90% confidence limit or $\Theta_{\mathrm{QCD}} < 10^{-10}$ [Pendlebury *et al.* (2015)].

Why is the strong CP violation parameter Θ_{QCD} so small? Usually, the value of Θ_{QCD} is taken as an external parameter in the theory just like the quark masses are. In this scenario, QCD alone offers no answer to this question. QCD symmetries allow for a possible Θ_{QCD} term but do not constrain its size. One possibility involves new light mass pseudoscalar particles called axions which would enter with tiny masses m_a and couplings to Standard Model particles suppressed by a single power of some large mass scale — the so called "axion decay constant". In this scenario, the vacuum expectation value of the axion field would act to cancel any strong CP violation associated with the Θ_{QCD} term [Peccei and Quinn (1977)] — this Peccei–Quinn mechanism is discussed in Section 5.3.4. Possible axions are discussed more in Chapters 5 and 11. There is a vigorous experimental programme aimed at looking for them. In other ideas, it has been suggested that QCD might self solve the strong CP problem with a vanishing Θ_{QCD} parameter induced by confinement dynamics [Nakamura and Schierholz (2023)]. Here, $\Theta_{\mathrm{QCD}} \rightarrow 0$ becomes an issue of renormalisation group flow in the infrared.

In any case, it is not sufficient to restrict QCD to perturbative gluon exchanges. In addition to inducing the large $\tilde{m}_{\eta_0}^2$ mass term for the η' non-perturbative gluon topology may play an important role in detailed confinement dynamics beyond just the issues associated with Gribov copies [Polyakov (1977)].

The tiny value of Θ_{QCD} might also be considered as a naturalness issue [Shifman (2020)] (why is this coupling so very suppressed relative to the other dimensionless gauge and Yukawa couplings?) in parallel to the Higgs mass and cosmological constant scales (why are they so small compared to the Planck scale despite quantum corrections which naively might push them towards M_{Pl}?). Any non-zero Θ_{QCD} value is linked to the quark masses and therefore to the Higgs sector of the Standard Model. The parameters of the Higgs sector including the particle masses might be determined at very high scales in connection with electroweak

Higgs vacuum stability, see Chapter 3. It is not implausible that the strong CP puzzle could be resolved by physics deep in the ultraviolet. The Higgs mass and cosmological constant scale hierarchy puzzles are discussed in Chapter 10.

In the ultraviolet, asymptotic freedom means that we can use perturbative QCD to describe high momentum transfer processes, typically with $Q^2 > 2$ GeV^2, where α_s is small enough to use perturbative methods. The 3 gluon vertex which gives us asymptotic freedom also leads to gluon bremsstrahlung resulting in gluon induced jets of hadronic particles which were first discovered in high energy e^-e^+ collisions at DESY [Ellis (2014)]. Experimental evidence for the number of QCD colours comes from the ratio of cross-sections for hadron to muon-pair production in high energy electron-positron collisions, $R_{e^-e^+}$ and the decay amplitude for $\pi^0 \to 2\gamma$. These observables link hadronic systems to photon initial/final states and are each proportional to the number of dynamical colours N_c giving an experimental confirmation of $N_c = 3$.

High energy deep inelastic scattering experiments probe the deep structure of hadrons by scattering high energy electron or muon beams off hadronic targets [Roberts (1990)]. Deeply virtual photon exchange acts like a microscope which allows us to look deep inside the proton. These experiments reveal a proton built of nearly free constituents, its quark and gluon partons, with quarks carrying the electric charge.

In high resolution scattering processes involving large momentum transfer squared, Q^2, hadrons dissolve into parton constituents with interactions driven by parton scattering processes. At leading order the structure functions (deep inelastic form-factors) measured in the cross-sections for these reactions factorise into the convolution of a flux of partons (parton distribution functions that describe the non-perturbative hadron structure) feeding into hard quark/gluon parton scattering processes which involve large momentum transfer and are described by perturbative QCD. Partons then hadronise as they leave the interaction region combining with small momentum quarks/gluons to give colour neutral hadrons in the final state. Examples are (semi-inclusive) deep inelastic scattering and high-energy proton-proton collisions. The factorisation theorem [Collins *et al.* (1989)] tells us that the parton distributions are process independent meaning that the parton distributions measured in deep inelastic scattering can be used to calculate processes in large momentum transfer hadron–hadron collisions. This is the so called twist two approximation. Higher twist $\mathcal{O}(1/Q^2)$ corrections measure more complicated quark-gluon correlations and lead to interesting effects in QCD spin processes [Aidala *et al.* (2013)]. They are typically small in deep inelastic kinematics with large Q^2.

The parton distribution functions measure the probability of finding quark and gluon partons with light-front momentum fraction x in the proton, $k_+^{\text{parton}} = x p_+^{\text{proton}}$, and, for polarised distributions, with a particular spin direction. This x is equal to the Bjorken variable $x_{\text{Bj}} = Q^2/2p.q$ in deep inelastic scattering with p the target momentum and q the momentum of the incident photon. Logarithmic scale dependence is observed in the evolution of quark and gluon parton distributions with increasing Q^2, just as predicted by perturbative QCD and asymptotic freedom. The parton distribution functions are most readily understood in terms of light-cone correlation functions involving the Fourier transfer along light-cone of point split hadronic matrix elements with the QCD theory formulated in the light-front gauge $A_+ = 0$ [Jaffe (1996); Llewellyn Smith (1989)]. As one probes deeper inside the proton with increasing Q^2 one becomes increasingly sensitive to parton radiation processes so that the proton's momentum is perceived as shared between increasingly more quark and gluon constituents. Under slow logarithmic QCD evolution the weight of the quark and gluon distributions is shifted to lower momentum fraction x with increasing Q^2. Quark and gluon parton scattering processes play a vital role in high energy hadronic collisions, e.g. at the Large Hadron Collider at CERN [Altarelli (2013a)]. Understanding precision QCD is essential to searches for new physics phenomena in these high energy collisions. A recent review is given in [Gehrmann and Malaescu (2022)].

Deep inelastic scattering experiments tell us that about 50% of the proton's momentum observed at high Q^2 is carried by gluons, consistent with the QCD prediction for the deepest structure of the proton. Deep inelastic scattering is also sensitive to the physics of low energy DChSB. One observes an excess of anti-down over anti-up quarks, $\int_0^1 dx (\bar{d} - \bar{u})(x) = 0.147 \pm 0.039$ at an average $Q^2 = 4$ GeV^2 [Arneodo *et al.* (1994)], the so-called Gottfried sum-rule violation. This result is expected from pion cloud fluctuations $p \rightarrow n\pi^+$ (with the extra down quark in the neutron and anti-down quark in the π^+) as well as Pauli blocking effects in the proton wavefunction, with about equal contributions expected from each effect and with just negligible contribution from perturbative QCD processes $g \rightarrow q\bar{q}$ [Melnitchouk *et al.* (1991)].

The spin structure of the proton brings together much of this phenomenology with sensitivity to many aspects of QCD physics and its symmetries. Polarised deep inelastic scattering experiments tell us that just about 30% of the proton's spin is carried by the spin of the quarks and antiquarks inside [Adolph *et al.* (2017)]. This result, originally a big surprise, comes from measurement of the first moment of the proton's spin structure function $g_1(x, Q^2)$, viz. $\int_0^1 dx g_1$ [Aidala *et al.* (2013)]. This moment enables us to extract the value of the proton's flavour singlet axial-charge $g_A^{(0)}$, the proton's forward matrix element of the flavour-singlet axial-vector current $J_{\mu 5} = \bar{u}\gamma_\mu\gamma_5 u + \bar{d}\gamma_\mu\gamma_5 d + \bar{s}\gamma_\mu\gamma_5 s$, viz. $2M_P s_\mu g_A^{(0)} = \langle p, s|J_{\mu 5}|p, s\rangle$ with $g_A^{(0)} \approx 0.3$. In a static quark model without quark motion one would obtain 100%. Confinement introduces a transverse momentum scale in the proton and hence a net quark orbital angular momentum contribution to its spin structure. Relativistic effects, the OGE colour hyperfine potential and the proton's virtual pion cloud induced by DChSB all shift angular momentum from quark helicity to orbital contributions [Bass and Thomas (2010)]. A positive gluon polarisation contribution in the proton further suppresses the value of $g_A^{(0)}$. Theoretical QCD analysis leads to the formula

$$g_A^{(0)} = \left(\sum_q \Delta q - 3\frac{\alpha_s}{2\pi}\Delta g\right)_{\text{partons}} + \mathcal{C}_\infty, \qquad (2.51)$$

see [Altarelli and Ross (1988); Bass (2005); Bass *et al.* (1991); Carlitz *et al.* (1988); Efremov and Teryaev (1988)]. Here, $\Delta g_{\text{partons}}$ is the amount of spin carried by polarised gluon partons in the polarised proton with $\alpha_s \Delta g \sim$ constant as $Q^2 \to \infty$ [Altarelli and Ross (1988); Efremov and Teryaev (1988)]. The growth in gluon polarisation at large Q^2 is compensated by similar growth with opposite sign in the gluon orbital angular momentum. The $\Delta q_{\text{partons}}$ measures the spin carried by quarks and antiquarks carrying "soft" transverse momentum $k_t^2 \sim \mathcal{O}(P^2, m^2)$, where P^2 is a typical gluon virtuality in the nucleon and m is the light quark mass. The polarised gluon term is associated with events in polarised deep inelastic scattering where the hard photon strikes a quark or antiquark generated from photon-gluon fusion and carrying $k_t^2 \sim Q^2$ [Carlitz *et al.* (1988)].[11] The term $\mathcal{C}_\infty$ denotes a potential non-perturbative gluon topological contribution with support only at Bjorken $x = 0$ [Bass (2005)]. It is associated with a possible subtraction constant in the dispersion relation for g_1 and, if finite, would be missed in polarised deep inelastic scattering experiments which kinematically measure $x_{\text{Bj}} > 0$ thus suppressing the value of $g_A^{(0)}$ extracted from these experiments.

---

[11]In connection with the flavour-singlet axial-vector current, this term is associated with the QCD axial anomaly in perturbative QCD, see Chapter 4.

One finds a consistent picture where valence quark dynamics, the pion cloud, modest gluon polarisation (up to about 50% of the proton's spin at the scale of typical deep inelastic experiments) and perhaps non-local gluon topology describe the internal spin structure of the proton [Aidala *et al.* (2013)]. For a recent discussion of proton spin dynamics see [Bass (2024)] and references therein.

2.3   The Standard Model and Higgs Phenomena

The electroweak part of the Standard Model is built from the gauge groups of $SU(2)_L$ (acting only on left-handed fermions) and $U(1)_Y$ hypercharge. After mixing between the charge neutral gauge bosons the physical gauge bosons become the massless photon of QED and the Z boson with mass 91 GeV. The charged $W^\pm$ bosons have mass 80 GeV. With the SU(2) gauge fields coupling just to left-handed fermions, parity is maximally violated in the weak sector.

Mass terms for gauge bosons violate gauge invariance without extra ingredients in the theory. This problem is resolved through the Brout–Englert–Higgs, BEH, mechanism which induces an accompanying scalar spin-zero Higgs boson. The underlying gauge symmetry is always present in the Lagrangian but becomes "hidden" in the ground state vacuum. In the Standard Model the BEH mechanism also gives mass to the charged fermions with these mass terms given by the couplings of the fermions to the BEH Higgs field.

The fermions come in three families of SU(2) electroweak doublets: the light up and down quarks plus electron and related neutrino with this structure repeated for charm and strange quarks and the muon plus its neutrino and also the heaviest top and bottom quarks plus the tau lepton and its neutrino. The down, strange and bottom quarks in the lower component of the SU(2) quark doublets mix through the 3×3 unitary Cabibbo–Kobayashi–Maskawa (CKM), matrix which also includes a CP violating phase factor. In the minimal Standard Model neutrinos have zero mass. Also, there are no (interacting) right handed neutrinos. Tiny but finite neutrino masses are necessary to explain neutrino oscillation measurements [Balantekin and Kayser (2018)] with the lightest neutrino mass expected to be $\approx 10^{-8}$ times the electron mass. Neutrinos are discussed in Section 5.3.1. A key open question is whether neutrinos might be their own antiparticles.

The 125 GeV mass Higgs boson discovered at CERN behaves very Standard Model like and completes the spectrum of the Standard Model.

With the observed boson the Standard Model has good ultraviolet behaviour (discussed in Chapter 3). It satisfies perturbative unitarity, is renormalisable and comes with an (almost) stable vacuum when the theory is extrapolated up to the Planck scale.

We next discuss the BEH mechanism first for the idealised case of massive U(1) gauge bosons and then its vital role in the electroweak Standard Model.

2.3.1   *Massive gauge bosons and the BEH mechanism*

When taken alone, mass terms for gauge bosons break the underlying gauge symmetry. For example, consider particles χ (fermions or scalar bosons) interacting with a spin-one gauge field A_ρ with the system invariant under the local gauge transformations $\chi \to e^{i\omega}\chi$ and $A_\rho \to A_\rho - \frac{1}{g}\partial_\rho\omega$. Here, ω is the phase parameter for the gauge symmetry, $\partial_\rho = \frac{\partial}{\partial x^\rho}$ is a partial derivative and g is the coupling of A_ρ to χ; ρ denotes the Lorentz index. Introducing a mass term $m^2 A_\rho A^\rho$ violates the gauge symmetry without extra ingredients.

This problem is resolved through the BEH mechanism developed in [Englert and Brout (1964); Higgs (1964a, 1964b, 1966)] and the related work [Guralnik *et al.* (1964); Kibble (1967)]. The gauge symmetry of the underlying theory can be hidden in the ground state. The symmetry parameter ω freezes out to a particular value with all possible values being degenerate. This spontaneous symmetry breaking process generates massless Nambu–Goldstone modes — one for each generator of the symmetry. For local gauge symmetries these massless Nambu–Goldstone modes combine with the gauge bosons to generate new longitudinal modes of the gauge fields and conserving the total number of degrees of freedom. The transverse and longitudinal components of the spin-one gauge field acquire non-zero mass, which is the same for both components. In addition, a new scalar boson is produced with finite coupling to the massive gauge fields — the Higgs boson.

To understand the BEH mechanism, consider the coupling of the gauge field A_ρ to a complex scalar field ϕ via the gauge covariant derivative with coupling constant g, namely $D_\rho\phi = [\partial_\rho + ig A_\rho]\phi$. Under the local gauge transformation $\phi \to e^{i\omega}\phi$, $D_\rho\phi \to e^{i\omega}D_\rho\phi$ with the partial derivative acting on ω compensated by the gauge transformation of A_ρ. The scalar field is taken with potential

$$V(\phi) = \mu^2\phi^2 + \lambda\phi^4. \tag{2.52}$$

Here, the self-coupling $\lambda \geq 0$ so that the potential has a finite minimum as required for vacuum stability. If $\mu^2 > 0$ the potential describes a particle with mass μ (not to be confused with the notation in previous discussion of renormalisation scales). When $\mu^2 < 0$ the potential is minimal at

$$|\phi| \equiv \frac{v}{\sqrt{2}} = \sqrt{-\frac{\mu^2}{2\lambda}}. \tag{2.53}$$

This potential is illustrated in the left panel of Fig. 2.2. Excitations around the degenerate minima of the potential — the bottom of the "Mexican hat" — correspond to a massless Nambu–Goldstone mode. Gauge freedom allows us to choose v as the vacuum expectation value (vev), of the real part of ϕ with all choices of vacuum states being degenerate and physically equivalent. Expanding the scalar field about this minimum of the potential, the Nambu–Goldstone mode is "eaten" to become the longitudinal mode of A_ν which now acquires mass $g^2 v^2$. The Higgs boson h with mass squared $m_h^2 = 2\lambda v^2$ corresponds to excitations up the rim of the potential. Quantum corrections to the Higgs potential — effects relevant at large energy scales and field values are discussed in Chapter 3.

The consistency of massive gauge bosons with gauge invariance was first solved by Anderson in the context of massive "photons", called plasmons, in superconductors [Anderson (1963)]. Here the photon behaves as a wave on a sea of Bardeen–Cooper–Schrieffer (BCS) Cooper pairs which in this case

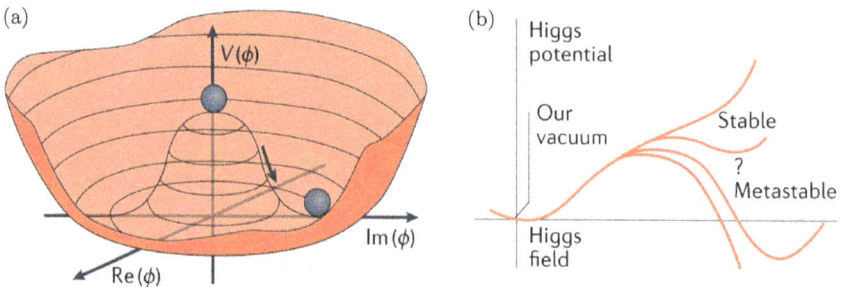

Fig. 2.2.   Left (a): The Higgs potential for $\mu^2 < 0$, Eq. (2.52). Choosing any of the points at the bottom of the potential spontaneously breaks the rotational U(1) symmetry. Right (b): Quantum corrections can change the shape of the Higgs potential as discussed in Chapter 3. Here, the minimum of "our vacuum" is taken at $|\phi| = \frac{v}{\sqrt{2}}$. Figure from [Bass et al. (2021)].

act as the scalar field ϕ, condensing in the ground state. The order parameter is not rigid with zero momentum Cooper pairs, but fluctuates in the longitudinal component to preserve the translational symmetry of the electron gas. The plasmon's transverse component is a modification of a real photon propagating in the plasma whereas the longitudinal mode is an attribute of the system. Massive plasmons are manifest through the exponential decrease of the magnetic field inside the superconductor (the Meissner effect).

The extension of this physics to relativistic dynamics [Englert and Brout (1964); Higgs (1964a, 1964b, 1966)] has been introduced to provide a consistent model of weak interactions in particle physics [Glashow (1961); Salam (1968); 't Hooft and Veltman (1972); Weinberg (1967)]. Contrary to the BCS case, the weak interaction requires the introduction of an additional fundamental scalar field. A dynamic explanation of the Higgs mechanism à la BCS would be a major breakthrough and is one of the fundamental motivations to measure with the highest possible precision the properties of the Higgs particle.

2.3.2 *The Standard Model*

In the Standard Model, the left-handed fermions transform under local SU(2) rotations and interact with the corresponding gauge bosons. Left- and right-handed fermions both transform under an extra "hypercharge" U(1) gauge symmetry. That is, the fermions transform under the SU(2) and U(1) gauge transformations as

$$\Psi_L(x) \rightarrow e^{i\frac{1}{2}\vec{\tau}\cdot\vec{\alpha}(x)}\Psi_L(x) \quad \text{and} \quad \Psi(x) \rightarrow e^{i\frac{y}{2}\beta(x)}\Psi(x), \tag{2.54}$$

where y is called the particle's hypercharge which is different for different types of particles, see Table 2.1. Restricting to one family of fermions, the left-handed fermions are described by lepton and quark doublets

$$\Psi = \text{L} = \begin{pmatrix} \nu \\ e \end{pmatrix}, \quad \Psi = \text{Q} = \begin{pmatrix} u \\ \tilde{d} \end{pmatrix}_c. \tag{2.55}$$

Here, u is the up quark and $\tilde{d}$ is the down quark component taking into account d-quark flavour mixing with the Cabibbo angle and 3×3 CKM matrix. The subscript c denotes the three colours of quarks (red, green and blue) associated with QCD.

Table 2.1. Standard Model fermions with their weak isospin (t, t_3), hypercharge y and electric charge $Q = t_3 + y/2$. Here $(\tilde{d}, \tilde{s}, \tilde{b})_L$ denote the lower components of the left-handed quark doublets after flavour mixing via the 3×3 Cabibbo–Kobayashi–Maskawa matrix.

		t	t_3	y	$Q = t_3 + y/2$
Doublets	$(\nu_l)_L$	1/2	1/2	−1	0
	l_L^-	1/2	−1/2	−1	−1
	$(u, c, t)_L$	1/2	1/2	1/3	2/3
	$(\tilde{d}, \tilde{s}, \tilde{b})_L$	1/2	−1/2	1/3	−1/3
Singlets	$(\nu_l)_R$	0	0	0	0
	l_R^-	0	0	−2	−1
	$(u, c, t)_R$	0	0	4/3	2/3
	$(d, s, b)_R$	0	0	−2/3	−1/3

Corresponding to the SU(2) and U(1) local gauge symmetries, let W_μ^i denote the SU(2) bosons and B_μ denote the U(1) gauge boson. The gauge covariant derivatives read

$$D_\mu \Psi_L = \left[\partial_\mu + \frac{1}{2} i g \vec{\tau}.\vec{W}_\mu + \frac{1}{2} i g' y B_\mu \right] \Psi_L$$

$$D_\mu \Psi_R = \left[\partial_\mu + \frac{1}{2} i g' y B_\mu \right] \Psi_R.$$

(2.56)

Here, τ denotes the SU(2) Pauli matrices and g and g' are the SU(2) and U(1) couplings. The electric charge is $Q = t_3 + y/2$ where $t_3 = \tau_3/2$. The hypercharge y is $y = -1$ for left-handed leptons l_L, $y = -2$ for the right-handed leptons l_R, $y = \frac{1}{3}$ for the left-handed quarks, $y = -\frac{2}{3}$ for right-handed down-type quarks, and $y = \frac{4}{3}$ for right-handed up-type quarks. These assignments ensure that the electron carries electric charge and the neutrino is electric charge neutral.

QED and weak interactions unify through mixing between the photon and the electric-charge zero Z boson with the gauge group SU(2)$_L \otimes$U(1). The massless photon A_μ and neutral Z boson denoted by the field Z_μ are linear combinations of the neutral weak SU(2) gauge boson and the U(1) hypercharge gauge boson, viz.

$$\begin{pmatrix} W_\mu^3 \\ B_\mu \end{pmatrix} = \begin{pmatrix} \cos\theta_W & \sin\theta_W \\ -\sin\theta_W & \cos\theta_W \end{pmatrix} \begin{pmatrix} Z_\mu \\ A_\mu \end{pmatrix}.$$

(2.57)

The mixing angle θ_W is called the Weinberg angle. The neutral current gauge bosons couple to the left-handed fermions through the interaction term

$$i\bar{\Psi}_L\gamma^\mu\left[\left(gt_3\sin\theta_W + g'\frac{y}{2}\cos\theta_W\right)A_\mu + \left(-g'\frac{y}{2}\sin\theta_W + g\cos\theta_W t_3\right)Z_\mu\right]\Psi_L$$

(2.58)

which becomes

$$ig\sin\theta_W\bar{\Psi}_L\gamma^\mu\left[\left(t_3 + \frac{1}{2}y\right)A_\mu + \left(-\frac{1}{2}y\tan\theta_W I + \cot\theta_W t_3\right)Z_\mu\right]\Psi_L$$

(2.59)

when we identify

$$g\sin\theta_W = g'\cos\theta_W = e$$

(2.60)

with e the electric charge. The ratio $\tan\theta_W = g'/g$. With $y = -1$ the neutrino has no electric charge and no coupling to the photon. Mixing fixes the Weinberg angle θ_W

$$\cos\theta_W = \frac{g}{\sqrt{g^2 + g'^2}}, \quad \sin\theta_W = \frac{g'}{\sqrt{g^2 + g'^2}},$$

(2.61)

i.e.

$$W_\mu^\pm = \frac{1}{\sqrt{2}}(W_\mu^1 \mp W_\mu^2), \quad B_\mu = \frac{-g'Z_\mu + gA_\mu}{\sqrt{g^2 + g'^2}}, \quad W_\mu^3 = \frac{gZ_\mu + g'A_\mu}{\sqrt{g^2 + g'^2}}.$$

(2.62)

The $W^\pm$ connect different members of the electroweak lepton and quark doublets whereas the photon and Z^0 are electric-charge neutral bosons.

The BEH mechanism gives mass to the W and Z gauge bosons while respecting the underlying gauge symmetries. For the Standard Model one has a scalar Higgs doublet Φ with potential

$$V(\Phi) = \mu^2|\Phi|^2 + \lambda|\Phi|^4.$$

(2.63)

The Higgs doublet transforms as

$$\Phi(x) \to e^{i\frac{1}{2}\vec{\tau}\cdot\vec{\alpha}(x)}\Phi(x) \quad \text{and} \quad \Phi(x) \to e^{i\frac{y}{2}\beta(x)}\Phi(x)$$

(2.64)

under the SU(2) and U(1) gauge transformations. The Higgs scalar doublet comes with the covariant derivative coupling

$$D_\mu \Phi = \left[\partial_\mu + \frac{1}{2} ig\vec{\tau}.\vec{W}_\mu + \frac{1}{2} ig' y_\phi B_\mu \right] \Phi. \tag{2.65}$$

Here, the scalar hypercharge is $y_\phi = +1$ for the Standard Model to ensure that the photon does not couple to the Higgs boson and that ϕ^+ has the correct charge. In the BEH potential, Eq. (2.63), $\lambda \geq 0$ so that $V(\Phi)$ has a finite minimum as required for electroweak vacuum stability. If $\mu^2 > 0$ Eq. (2.63) describes the potential for a particle with mass μ. The interesting case relevant to the Standard Model is with $\mu^2 < 0$. In this case, the potential is minimal at

$$|\Phi| = \frac{v}{\sqrt{2}} \equiv \sqrt{-\frac{\mu^2}{2\lambda}}, \tag{2.66}$$

where v is the vacuum expectation value, vev, of the BEH field. One takes the vev to be real,

$$\langle \Phi \rangle = \frac{1}{\sqrt{2}} \begin{pmatrix} 0 \\ v \end{pmatrix}, \tag{2.67}$$

and expands the scalar field about this vev, viz.

$$\Phi = \begin{pmatrix} \phi^+ \\ \frac{1}{\sqrt{2}}(v + \rho + i\zeta) \end{pmatrix}. \tag{2.68}$$

The phase of Φ is then chosen using the gauge freedom to make Φ real. The Standard Model is most transparent when formulated in this so-called unitary gauge. Out of the Higgs doublet the three massless Nambu–Goldstone modes associated with spontaneous symmetry breaking decouple. The ζ and $\phi^\pm$ components are "eaten" to become the longitudinal modes of the now massive W and Z bosons, thus conserving the number of degrees of freedom. The fourth component of the BEH field Φ, $h = \rho$, is the scalar Higgs particle. The mass of this Higgs particle is related to the Higgs self-coupling parameter λ by $m_h^2 = 2\lambda v^2$. The spontaneous symmetry breaking is defined relative to the choice of gauge, e.g. the unitary gauge, with all gauge choices being physically equivalent [Kibble (2014)]. The unitary gauge is commonly used at tree level whereas the Feynman–'t Hooft gauge is used in loop diagram calculations where it simplifies the calculations.

The term involving the covariant derivative of Φ follows from Eq. (2.65), resulting in

$$|D_\mu \Phi|^2 = \frac{1}{2}|\partial_\mu h|^2 + \left[\frac{g^2 v^2}{4} W_\mu^+ W^{-\mu} + \frac{(g^2 + g'^2)\, v^2}{8} Z_\mu Z^\mu \right] \left(1 + \frac{h}{v} \right)^2.$$

$$(2.69)$$

From this expression the masses of the W and Z bosons can be determined as follows

$$m_W = \frac{gv}{2}, \quad m_Z = \frac{\sqrt{g^2 + g'^2}\; v}{2}, \qquad (2.70)$$

i.e. with

$$m_W = m_Z \cos\theta_W = \frac{1}{2}gv. \qquad (2.71)$$

The photon remains massless with $m_A = 0$ corresponding to a vanishing coefficient of the linear combination proportional to $g'W_\mu^3 + gB_\mu$. The Weinberg angle is accurately measured with $\sin^2\theta_W(M_Z)|_{\overline{\text{MS}}} = 0.23129(4)$ [Navas *et al.* (2024)].

With $m_h^2 = 2\lambda v^2$ the Higgs potential becomes

$$V(\Phi) = \frac{1}{2}m_h^2 h^2 + \frac{m_h^2}{2v}h^3 + \frac{m_h^2}{8v^2}h^4 - \frac{1}{8}m_h^2 v^2 \qquad (2.72)$$

including the mass term for the Higgs boson and the 3- and 4-Higgs boson interactions. One also has the constant, field independent, term $\frac{1}{8}m_h^2 v^2$ which decouples from Standard Model dynamics but does provide a contribution to the net vacuum energy measured through the cosmological constant, see Chapters 9 and 10.

The BEH mechanism also generates the masses of quarks and charged leptons. Singlet mass terms are constructed by contracting the left-handed fermion doublets with the SU(2) Higgs doublet, including the vev, and then coupling to the right-handed fermion singlets, viz.

$$\mathcal{L}_Y = -y_d \bar{Q}_L \Phi d_R - y_u \bar{Q}_L \tilde{\Phi} u_R - y_l \bar{L}_L \Phi l_R \; + \; \text{h.c.} \qquad (2.73)$$

which for the first generation of light fermions gives

$$\mathcal{L}_Y = -\left(1 + \frac{h}{v} \right)\{ m_d \bar{d}d + m_u \bar{u}u + m_l \bar{l}l \}. \qquad (2.74)$$

The resulting fermion masses are proportional to the Yukawa couplings of the fermions to the Higgs boson. In the Standard Model, the left-handed fermion doublets mean that the Higgs doublet is required for fermion masses. QED type mass terms are not possible. Without the Higgs there is no bare fermion mass term in the Standard Model [Veltman (1997)].[12] Whereas the gauge boson masses change the degrees of freedom with longitudinal polarisation of the gauge bosons becoming physical, fermion masses are a continuous deformation of the massless theory.

The Standard Model particle masses and couplings are related. For the W and Z gauge bosons

$$m_W^2 = \frac{1}{4}g^2 v^2, \quad m_Z^2 = \frac{1}{4}(g^2 + g'^2)v^2. \tag{2.75}$$

The charged fermion masses are

$$m_f = y_f \frac{v}{\sqrt{2}} \quad (f = \text{quarks and charged leptons}), \tag{2.76}$$

where y_f are the Yukawa couplings and the Higgs boson mass is

$$m_h^2 = 2\lambda v^2 \tag{2.77}$$

with λ the Higgs self-coupling. Before considering neutrinos, the Standard Model has 18 parameters: 3 gauge couplings and 15 associated with the Higgs sector (6 quark masses, 3 charged lepton masses, 4 quark mixing angles including one CP violating complex phase, the W and Higgs boson masses).

There is a wide range of particle masses with $m_W = 80$ GeV, $m_Z = 91$ GeV, $m_h = 125$ GeV and the charged fermion masses ranging from 0.5 MeV for the electron up to 173 GeV for the top quark. The charged lepton masses are $m_e = 0.51$ MeV, $m_\mu = 106$ MeV and $m_\tau = 1777$ MeV. The down and up type quarks come with masses $m_d = 5$ MeV, $m_s = 93^{+8}_{-3}$ MeV, $m_b = 4.78 \pm 0.06$ GeV and $m_u = 2$ MeV, $m_c = 1.67 \pm 0.07$ GeV, $m_t = 172.69 \pm 0.30$ GeV with the light u, d, s quark masses quoted in the $\overline{\text{MS}}$ scheme at the renormalisation scale $\mu = 2$ GeV and the heavy c, b, t masses quoted as their Particle Data Group, PDG, pole masses [Navas *et al.* (2024)].

---

[12]Actually, mass is important even in QED where charged fermion masses are essential for consistency [Gómez and Letschka (2020); Gribov (1982); Morchio and Strocchi (1986)]. Due to infrared divergences one cannot perturbatively renormalise QED on-shell with massless charged fermions. If one renormalises the massive theory on-shell and then takes the mathematical limit that the electron mass goes to zero in Eq. (2.8), then the QED Landau pole gets pulled into the infrared. The chiral coupling of left handed fermions to SU(2) gauge bosons in the Standard Model means that quarks and charged leptons get their masses through the BEH mechanism.

The value of the Higgs vacuum expectation value v can be determined by matching the Standard Model to the Fermi four-fermion interaction theory of weak interactions at low energies $E \ll v$. This gives the relation

$$\frac{G_F}{\sqrt{2}} = \frac{g^2}{8m_W^2} = \frac{1}{2v^2}, \tag{2.78}$$

where the Fermi coupling constant $G_F \simeq 1.166 \times 10^{-5}\ \text{GeV}^{-2}$. Numerically, one finds

$$v \simeq (\sqrt{2}G_F)^{-\frac{1}{2}} = 246\ \text{GeV}. \tag{2.79}$$

It follows that $\lambda = m_h^2/2v^2 \simeq 0.13$ with the Higgs mass $m_h = 125\ \text{GeV}$ measured at the LHC. In the Standard Model, once the Higgs boson mass is known the interactions of all vector bosons with the Higgs boson are determined as well as the Higgs boson self-interactions through the triple and quartic Higgs boson couplings.

The Standard Model relations in Eqs. (2.75) and (2.76) have been tested to about 10% precision for the W and Z gauge bosons, the heavy top and bottom quarks, and the τ and μ charged leptons. The measured coupling strengths scale as a function of the particle masses just as predicted by the Standard Model, see Fig. 2.3 and the original references [Aad *et al.* (2021, 2022); Sirunyan *et al.* (2021)]. This figure shows the masses scaled by so-called "coupling modifiers" κ applied to the Standard Model couplings of the Higgs boson to other particles: κ_V for the coupling of the Higgs boson to vector gauge bosons and κ_F for the couplings to fermions. The coupling modifiers are derived from global fits to all the measurements in different production and decay channels assuming Standard Model relations between the channels. These κ values therefore do not measure the couplings directly but show levels of deviation from the Standard Model expectations. We can test the size of these coupling-strength modifiers and the values come out consistent with one, viz. the Standard Model prediction. A direct measurement of λ awaits future collider experiments with higher energy and higher luminosity [Jakobs and Zanderighi (2023)].

With three families of fermions, the lower components of the quark doublets mix through a unitary 3×3 matrix — the Cabibbo–Kobayashi–Maskawa, CKM, matrix [Cabibbo (1963); Kobayashi and Maskawa (1973)]. This matrix contains three real mixing parameters and one CP violating angle. (Three families are the minimum needed to give a CP violating phase in the mixing matrix [Jarlskog (1985)].) The measured mixing parameters from the light d and s quark mixing with the Cabibbo angle extended to

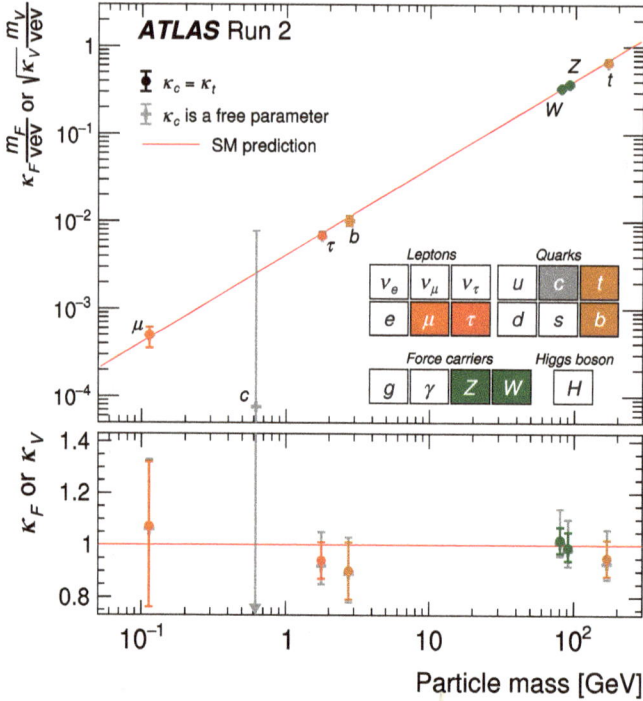

Fig. 2.3. Higgs boson coupling strength modifiers and their uncertainties versus the particle masses as measured by the ATLAS collaboration and compared to the Standard Model prediction. Loop-induced processes are resolved according to the Standard Model predictions, and no decays into non Standard Model particles are allowed. The lower panel shows the values and uncertainties of the modifiers κ_V and κ_F. Figure from [Aad *et al.* (2022)].

the 3×3 CKM matrix are consistent with a unitary matrix, without need for extra ingredients.[13]

---

[13]The unitary CKM matrix describing the quark flavour mixing is

$$V_{\text{CKM}} = \begin{pmatrix} 1 & 0 & 0 \\ 0 & c_{23} & s_{23} \\ 0 & -s_{23} & c_{23} \end{pmatrix} \begin{pmatrix} c_{13} & 0 & s_{13}e^{-i\delta} \\ 0 & 1 & 0 \\ -s_{13}e^{-i\delta} & 0 & c_{13} \end{pmatrix} \begin{pmatrix} c_{12} & s_{12} & 0 \\ -s_{12} & c_{12} & 0 \\ 0 & 0 & 1 \end{pmatrix}, \quad (2.80)$$

where $c_{ij} = \cos\theta_{ij}$ and $s_{ij} = \sin\theta_{ij}$ connect different quark states with θ_{12} the Cabibbo angle. Numerically, one finds $\sin\theta_{12} = 0.22501 \pm 0.00068$, $\sin\theta_{13} = 0.00373 \pm 0.00009$, $\sin\theta_{23} = 0.04183^{+0.00079}_{-0.00069}$ and $\delta = 1.147 \pm 0.026$ with the CKM matrix consistent with unitarity [Navas *et al.* (2024)].

The high energy limit of the Standard Model is discussed in Chapter 3.

The present experimental status of the Standard Model is reviewed in [Altarelli (2013a); Schael *et al.* (2006)] covering previous precision studies using the LEP and SLC experiments. The LHC results from ATLAS are covered in [Aad *et al.* (2024c)] and CMS in [Hayrapetyan *et al.* (2024b)]. Top quark studies are covered in the ATLAS [Aad *et al.* (2024b)] and CMS [Hayrapetyan *et al.* (2024a)] reports. Higgs measurements are reviewed in the reports from ATLAS [Aad *et al.* (2022, 2024a)] and CMS [Tumasyan *et al.* (2022)]. Further recent reviews of Higgs physics are given in [Bass *et al.* (2021)] and [Jakobs and Zanderighi (2023)]. The comparison of measured LHC cross sections with Standard Model theory predictions is shown in Fig. 2.4.

If the Standard Model is extended to include tiny neutrino masses suggested by neutrino oscillations, then the number of neutrino parameters depends on whether neutrinos are Dirac or Majorana fermions. (Majorana fermions are their own antiparticles.) If neutrinos are Dirac particles, then their right-handed states would be "sterile" in the Standard Model

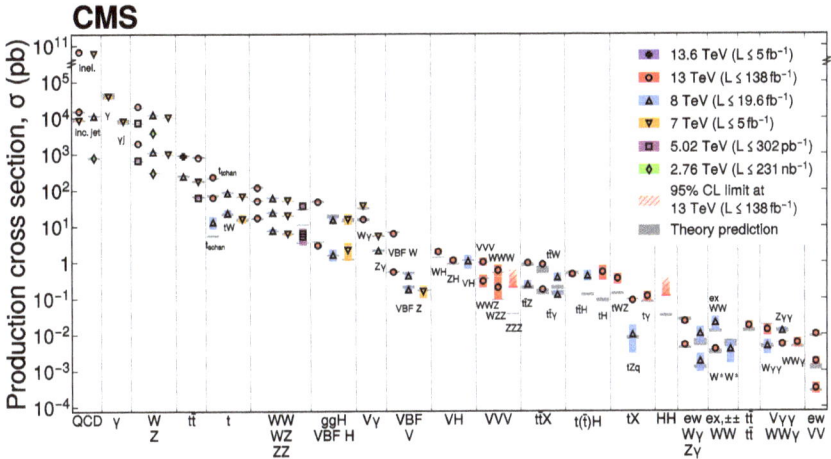

Fig. 2.4. Cross sections of selected high-energy processes measured by the CMS experiment. Measurements performed at different LHC proton–proton collision energies are marked by unique symbols and the coloured bands indicate the combined statistical and systematic uncertainty of the measurement. Grey bands indicate the uncertainty of the corresponding Standard Model theory predictions. Shaded hashed bars indicate the excluded cross section region for a production process with the measured 95% CL upper limit on the process indicated by the solid line of the same colour. Figure from [Hayrapetyan *et al.* (2024b)].

with no coupling to the Standard Model gauge bosons. For Dirac neutrinos one has 7 parameters associated with neutrino mixing: 3 masses plus 4 angles mirroring the quark CKM matrix. With Majorana neutrinos, one finds 9 neutrino parameters (3 masses plus 6 mixing angles including 3 CP violating phases). For Majorana neutrinos the neutrino masses might be connected to a dimension 5 operator, suppressed by a single power of some large mass scale $\approx 10^{16}$ GeV and thus sensitive to physics at very high energies [Weinberg (1979)]. The generalisation of the CKM matrix to neutrinos is known as the Pontecorvo–Maki–Nakagawa–Sakata, PMNS, matrix [Maki *et al.* (1962); Pontecorvo (1957)]. Neutrinos are discussed in Section 5.3.1.

The BEH mechanism comes with a subtlety. Elitzur's theorem tells us that there is no gauge invariant order parameter associated with the BEH mechanism (since gauge symmetry does not act on physical states) [Elitzur (1975)]. Spontaneous symmetry breaking is defined relative to the choice of gauge, e.g. the unitary gauge, with all gauge choices being physically equivalent.

There is a related issue with the definition of asymptotic states. Gauge symmetries are properties of internal degrees of freedom with the physical Hilbert space gauge invariant. This means that gauge symmetries do not connect physical particle states. The physical particles in the Standard Model (fermion asymptotic states) turn out to be SU(2) gauge singlets with gauge degrees of freedom transported away by Higgs ghosts. Particle states defined via gauge singlet composite operators involving the fermion doublets and gauge bosons plus the Higgs doublet were considered in ['t Hooft (1980a)] in the context of a "confinement phase" model as well as in [Frohlich *et al.* (1980, 1981)], where the propagators corresponding to the gauge singlet composite operators were found to be coincident with perturbative Standard Model ones at leading order in the Higgs vacuum expectation value. Compositeness here has nothing to do with real bound states and binding energy. The decoration Higgs-ghost fields just produce singlet fields while the Higgs-ghosts are transporting away the gauge degrees of freedom. In general, the physics behaves just as in unitary gauge (where Higgs ghosts are absorbed as the longitudinal components of the W and Z bosons). These fields are gauge orbits of the unitary gauge fields and do not change the physics at all as we know it. The bookkeeping within a calculation looks very different though; for detailed discussion see [Jegerlehner and Fleischer (1985, 1986)] and [Banks and Rabinovici (1979)].[14]

The Universe is very sensitive to values of Standard Model parameters and hence to the Higgs couplings. The parameters of particle physics interactions are closely correlated to the conditions for our existence [Carr and Rees (1979); Livio and Rees (2005)]. Small changes in the Higgs couplings and particle masses can lead to a very different Universe assuming that the vacuum remains stable. One example is that small changes in the light quark masses can prevent Big Bang nucleosynthesis [Carr and Rees (1979)]. Once radiative corrections are taken into account the stability of the Higgs vacuum is very sensitive to the values of the top quark and Higgs boson masses. This means that the world of everyday experience built out of first generation light fermions $u, d, e, \bar{\nu}_e$ is actually very dependent on the properties of these heavy particles to ensure that the lightest particles are defined relative to a stable vacuum.

2.3.3 *CP violation and the electron EDM*

Searches for new sources of CP violation are motivated by the matter-antimatter asymmetry in the Universe which requires some extra source of CP violation beyond the quark mixing described by the CKM matrix in the electroweak Standard Model.

Precision measurements of particle electric dipole moments, EDMs, are an important probe of possible CP violation complementary to collider experiments [Comparat *et al.* (2023)]. For example, we have seen above the neutron EDM as a probe of the strong CP parameter Θ_{QCD}. An important observable is the electron's electric dipole moment, eEDM, d_e which is a measure of possible CP violation from any new physics induced anisotropy in the vacuum polarisation of the electron. Is the electron round? One finds that any d_e is tiny. The most accurate measurement of the eEDM comes

---

[14]When we perform local SU(2) gauge rotations of the fermion doublets we also gauge rotate the Higgs doublet as well as rotating around the bottom of the BEH Mexican hat potential in the vacuum. The fermions are quantised as excitations above this vacuum with all gauge choices physically equivalent, e.g. to the situation in the unitary gauge. Relative to the vacuum potential, including gauge rotations, nothing changes. The physical particles are singlets with respect to the local SU(2) transformations.

from the JILA experiment [Roussy *et al.* (2023)],

$$|d_e| < 4.1 \times 10^{-30} e\text{cm}, \qquad (2.81)$$

or $|d_e| < 2.1 \times 10^{-29} e\text{cm}$ when the JILA results are combined with results from the previous best ACME experiment at Harvard [Andreev *et al.* (2018)]. These experiments put strong limits on new CP violating interactions coupling to the electron. Within typical extensions of the Standard Model involving possible new heavy particles, the eEDM measurements constrain the mass scales of such interactions to energies greater than about the energy of the LHC. In the pure Standard Model, the eEDM prediction is $\approx 10^{-38} e\text{cm}$ so measurement of anything bigger would signify some new physics.

Chapter 3

Vacuum Stability and Running Couplings

If taken as the Standard Model Higgs boson, the 125 GeV boson discovered at the LHC completes the spectrum of the Standard Model. It comes accompanied by intriguing observations including possible clues to deeper new physics. Up to what energy might the Standard Model work as the theory of particle interactions? That is, what is the ultraviolet limit of the Standard Model when taken as an effective theory? What new interactions might lie beyond the Standard Model? Important theoretical issues include the stability of the Higgs vacuum and the hierarchy or naturalness puzzle: the small size of the electroweak scale and Higgs boson's mass relative to the Planck scale, $M_{Pl} = 1.2 \times 10^{19}$ GeV where quantum gravity effects might apply. In this chapter, we explain the issue of Higgs vacuum stability and why this might be hinting at new critical phenomena in the ultraviolet. Naturalness issues are discussed in Chapter 10.

With the discovered Higgs boson, the Standard Model is perturbative and predictive when it is extrapolated up to very high energies. Perturbative unitarity and the stability of the vacuum are each sensitive to the values of the Standard Model's parameters. The parameters measured in experiments and the ultraviolet behaviour of the Standard Model are thus strongly correlated within this theory.

Besides ensuring gauge invariance with the massive W and Z gauge bosons, the BEH mechanism also plays a vital role in ensuring consistent high energy behaviour of scattering amplitudes. The Higgs boson with mass 125 GeV guarantees the perturbative unitarity of high energy collisions involving massive W and Z bosons. The Higgs boson cancels terms from the longitudinal component of the W and Z bosons that would otherwise violate perturbative unitarity meaning scattering probabilities calculated using

Feynman diagrams would grow larger than one [Bell (1973); Cornwall *et al.* (1973, 1974); Llewellyn Smith (1973)]. Unitarity in high energy collisions involving massive spin-one bosons beyond simple U(1) theories requires both Yang–Mills structure plus the BEH mechanism [Llewellyn Smith (1973)]. The Higgs boson is also essential for the renormalisability of the theory, namely to ensure a consistent treatment of the ultraviolet divergences which appear in Feynman diagrams involving loops ['t Hooft (1971a); 't Hooft and Veltman (1972); Veltman (1968)]. The Higgs boson cannot be too heavy to do its job of maintaining perturbative unitarity. If the Higgs boson had not been found at the LHC some alternative mechanism would have been needed in the energy range of the experiments, for example involving strongly interacting W^+W^- scattering with the Higgs boson replaced by some broad resonance in the WW system [Chanowitz (2005)]. The Higgs boson discovered at CERN so far behaves very Standard Model like in all its measured properties.

Renormalisation group evolution of the Standard Model to high energy scales needs as input the couplings and particle masses

$$m_W^2 = \frac{1}{4}g^2v^2, \quad m_Z^2 = \frac{1}{4}(g^2 + g'^2)v^2, \quad m_f = y_f\frac{v}{\sqrt{2}}, \quad m_h^2 = 2\lambda v^2.$$

$$(3.1)$$

As shown in Fig. 2.3 these Standard Model relations have been tested to good precision about 10% for the W and Z gauge bosons, the heavy top and bottom quarks, and the τ and μ charged leptons.

For renormalisation group evolution one also needs input on the Higgs self-coupling λ. In the absence of direct measurements we have to assume the Standard Model relation connecting λ to the Higgs mass in Eq. (2.77). Taking $v = 246$ GeV and $m_h = 125$ GeV gives $\lambda = m_h^2/2v^2 \approx 0.13$ at the laboratory energy scale of the LHC. Measurement of λ awaits future collider experiments with a precision of about 50% expected from the high luminosity upgrade of the LHC and 5% precision requiring a new 100 TeV higher energy collider [Jakobs and Zanderighi (2023)].

Given these Standard Model couplings we can extrapolate the theory up to very high energy scales using perturbative renormalisation group equations. If we assume no coupling to undiscovered new particles, then it remains finite and well behaved with no Landau pole singularities below the Planck scale. That is, the Standard Model is mathematically consistent up to the Planck scale. Further, the Standard Model Higgs vacuum comes out very close to the border of stable and metastable, within about 1.3 σ of being stable [Bednyakov *et al.* (2015)] — see also [Alekhin *et al.* (2012); Bezrukov *et al.* (2012); Buttazzo *et al.* (2013); Degrassi *et al.* (2012); Jegerlehner (2014c); Masina (2013)]. Vacuum stability is connected with the renormalisation group running of the Higgs self-coupling λ with the vacuum stable for positive definite λ. If this coupling crosses zero deep in the ultraviolet and below the Planck scale, then it can lead to vacuum metastability. Whether this happens is very sensitive to the values of the Higgs boson and top quark masses and to details of calculations of higher order radiative corrections. In the case of a metastable vacuum there would be a second minimum in the Higgs potential at a lower value than that measured at our energy scale after inclusion of quantum radiative corrections. For an unstable vacuum the BEH potential would become unbounded from below at large values of the BEH field. These scenarios are sketched in Fig. 2.2(b).

Calculations are performed with the Standard Model evolved up to the Planck scale with the measured masses and couplings as input and using three loop renormalisation group, two loop matching plus pure QCD corrections evaluated to 4 loops. (Matching schemes are discussed in [Jegerlehner *et al.* (2013)].) In general, a higher top quark mass tends to reduce λ deep in the ultraviolet whereas a larger Higgs mass tends to increase it. Sensitivity to QCD corrections involving α_s means sensitivity also to the numbers of colours and active flavours and to the QCD scale Λ_{QCD} with studies of QCD effects reported in [Hiller *et al.* (2024)]. Both electroweak and QCD physics thus enter the vacuum stability calculations.

The Standard Model U(1) and SU(2) couplings run with the renormalisation group scale, just like the running of α in QED and α_s in QCD. The leading-order β_i–function coefficients are

$$
b_i = \begin{pmatrix} b_1 \\ b_2 \\ b_3 \end{pmatrix} = \frac{1}{4\pi} \left\{ \begin{pmatrix} 0 \\ -22/3 \\ -11 \end{pmatrix} + N_{\text{Fam}} \begin{pmatrix} 4/3 \\ 4/3 \\ 4/3 \end{pmatrix} + N_{\text{Higgs}} \begin{pmatrix} 1/10 \\ 1/6 \\ 0 \end{pmatrix} \right\}.
$$
(3.2)

Here, the subscript $i = 1, 2, 3$ refers to U(1)$_Y$, SU(2)$_L$ and QCD respectively. We take the same normalisation for the b_i as for the β–functions of QED in Eq. (2.6) and QCD in Eqs. (2.26)–(2.28). In Eq. (3.2), N_{Fam} is the number of active fermion families (noting that there are two flavours per family when comparing with the QCD formula, Eq. (2.28)), and N_{Higgs} is the number of Higgs doublets (which is equal to one for the Standard Model). The derivation of these β–function coefficients and the extension to higher orders is discussed in [Davies *et al.* (2020); Jones (1982); Machacek and Vaughn (1983)].

Beyond the renormalisation group running of the Standard Model gauge couplings $\alpha_i = g_i^2/4\pi$ (or equivalently the g_i), the renormalisation group running of the top quark Yukawa coupling y_t is described by the β-function

$$
\beta(y_t) = \mu^2 \frac{d}{d\mu^2} y_t(\mu^2)
$$

$$
= \frac{1}{16\pi^2} \left(\frac{9}{2} y_t^3 - \frac{17}{12} g'^2 y_t - \frac{9}{4} g^2 y_t - 8g_s^2 y_t \right) + \cdots \qquad (3.3)
$$

Here one finds a vital role for QCD interactions via the large term involving g_s^2 in generating the negative net sign for the β-function for y_t [Bednyakov *et al.* (2013b); Chetyrkin and Zoller (2012); Machacek and Vaughn (1984)].

The β-function for the Higgs self-coupling

$$
\beta(\lambda) = \mu^2 \frac{d}{d\mu^2} \lambda(\mu^2)
$$

$$
= \frac{1}{16\pi^2} \left(4\lambda^2 - 3g'^2\lambda - 9g^2\lambda + 12y_t^2\lambda + \frac{9}{4}g'^4 + \frac{9}{2}g'^2g^2 + \frac{27}{4}g^4 - 36y_t^4 \right)
$$

$$
+ \cdots \qquad (3.4)
$$

[Bednyakov *et al.* (2013a); Chetyrkin and Zoller (2012, 2013); Ford *et al.* (1993); Machacek and Vaughn (1985)] is found to have a zero deep in the ultraviolet

$$
\beta(\lambda) \sim \lambda \sim 0. \qquad (3.5)
$$

The key issue for electroweak vacuum stability is whether λ crosses zero below the Planck scale. The sign of $\beta(\lambda)$ is dominated by the large negative top quark Yukawa coupling which yields a negative β-function at laboratory energies and remains negative up to (close to) the Planck scale. In the absence of the large top quark coupling (and also with the Yukawa sector switched off), the sign of $\beta(\lambda)$ would be driven by the positive contribution from λ (and the other bosons). It turns out that the running of λ is more sensitive to the value of y_t than to λ itself. Formally, the vanishing of $\beta(\lambda)$ is not a fixed point in that it depends on other couplings which also change with the scale.

The Standard Model running couplings are shown in Fig. 3.1 which displays calculations [Bass and Krzysiak (2020a)] performed using the C++ renormalsation group evolution package [Kniehl *et al.* (2016)].

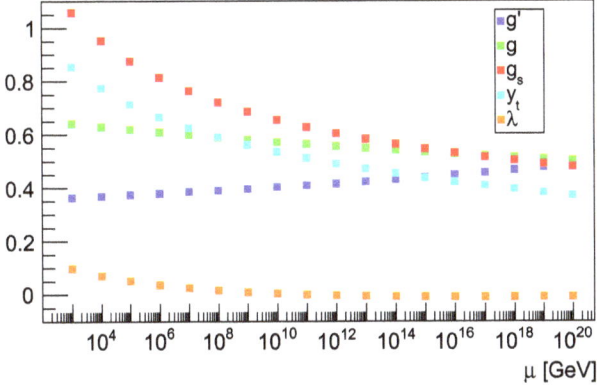

Fig. 3.1.   Running of the Standard Model gauge couplings g, g', g_s for the electroweak SU(2) and U(1) and colour SU(3), the top quark Yukawa coupling y_t and Higgs self-coupling λ. (From left, the points describe the evolution of g_s, y_t, g, g', λ in descending order.) Figure taken from [Bass and Krzysiak (2020a)].

The QCD and electroweak SU(2) couplings are asymptotically free, decaying logarithmically with increasing resolution, and the U(1) coupling is non-asymptotically free rising in the ultraviolet. These three couplings almost meet in the ultraviolet but don't quite. The top quark Yukawa coupling decreases with increasing resolution. With the measured top quark mass m_t and QCD coupling α_s the Standard Model needs a Higgs mass m_h bigger than about 125 GeV to ensure vacuum stability, making the Higgs particle discovered at CERN especially interesting. (The Higgs self coupling would generate a Landau pole singularity below the Planck scale if the Higgs mass were about 30% larger with the measured value of m_t, thus placing a perturbative upper bound on the possible Higgs mass [Hambye and Riesselmann (1997)].) To illustrate sensitivity to the top quark mass, if we set $m_h = 125$ GeV and vary the top quark mass in these calculations, then the Higgs self-coupling λ crosses zero around 10^{10} GeV with $m_t = 173$ GeV and remains positive definite up to M_{Pl} with $m_t = 171$ GeV. The recent and most accurate measurement of m_t from CMS Run-2 data collected at $\sqrt{s} = 13$ TeV is 171.77 ± 0.37 GeV [Tumasyan *et al.* (2023)].

Summaries of the most recent LHC top-quark measurements are given in the ATLAS [Aad *et al.* (2024b)] and CMS [Hayrapetyan *et al.* (2024a)] reports. The PDG value of the Higgs boson mass is 125.20 ± 0.11 GeV [Navas *et al.* (2024)] with LHC Higgs studies reviewed in the ATLAS [Aad *et al.* (2022, 2024a)] and CMS [Tumasyan *et al.* (2022)] reports.

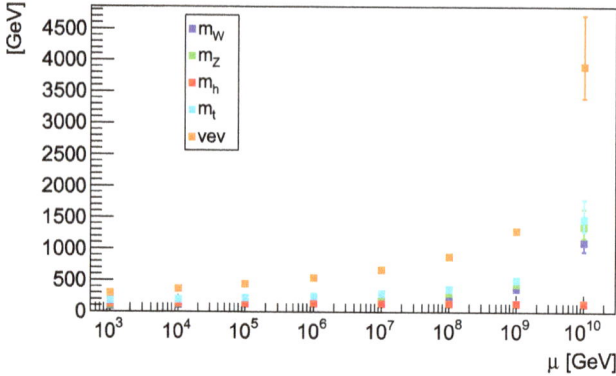

Fig. 3.2.   Running $\overline{\text{MS}}$ masses and the Higgs vev in the Standard Model. For the relation to the PDG pole masses, see [Kniehl *et al.* (2016)]. Uncertainties are calculated by varying all PDG values up and down by their respective uncertainties. Figure taken from [Bass and Krzysiak (2020a)].

Theoretical issues connecting the top quark Yukawa coupling and the measured mass are discussed in [Hoang (2020)]. Vacuum metastability here would correspond to a vacuum lifetime greater than $\sim 10^{600}$ years [Buttazzo *et al.* (2013)], which is very much bigger than the present age of the Universe of 13.8 billion years.

Corresponding to the running couplings in Fig. 3.1, Fig. 3.2 shows the renormalisation group behaviour of the Standard Model masses and Higgs vacuum expectation value v in the $\overline{\text{MS}}$ scheme. These masses and vacuum expectation value are related to the running couplings through Eqs. (2.75)–(2.77). The Higgs mass m_h evolves smoothly with increasing values of the scale μ. In Fig. 3.2, the divergence of v at $\mu \approx 10^{10}$ GeV corresponds to the scenario where λ crosses zero at this scale. With a fully stable vacuum λ remains positive and v is finite without any divergence.

Summarising, if we assume that there are no couplings to extra particles at higher energies, then the Standard Model revealed by current experiments is strongly correlated with its behaviour in the extreme ultraviolet. This may be telling us something deeper about the origin of the Standard Model. It is important to emphasise the large extrapolations in these calculations when evolving the Standard Model couplings up to the Planck scale. The existence of new physics, even at the largest scales, can affect the vacuum stability [Branchina and Messina (2013)]. Modulo this caveat, within the uncertainties on the top mass and radiative corrections, vacuum stability is certainly not excluded up to the energies where Grand Unified Theories are sometimes conjectured to apply or even up to the Planck scale [Bednyakov *et al.* (2015); Jegerlehner (2014c)]. The parameters of the Standard Model might thus be linked to physics at these high scales. In particular, the Higgs mass might be environmentally determined, linked to the stability of the electroweak vacuum.

The Higgs vacuum sitting "close to the edge" of stable and metastable may be pointing to possible new critical phenomena in the ultraviolet [Buttazzo *et al.* (2013); Degrassi *et al.* (2012); Jegerlehner (2014c)]. One interpretation is a statistical system in the ultraviolet close to the Planck scale and close to its critical point. There are ideas that near-criticality might act as an attractor point in the dynamical evolution of a higher energy phase, with analogous systems in Nature discussed in [Buttazzo *et al.* (2013)]. With just small changes in the Standard Model parameters the emerging low energy theory would be very different from the Standard Model.

If the Standard Model might really work up to these large scales, then how might we understand the origin of gauge symmetries and the new physics needed to the explain the open puzzles at the interface of particle physics and cosmology? Following Chapter 4 on renormalisation and ultraviolet completion issues, the rest of this book discusses this issue with emphasis on the idea that the Standard Model might be emergent below a scale deep in the ultraviolet, perhaps about 10^{16} GeV.

Chapter 4

Renormalisation and Quantum Field Theory Anomalies

The Standard Model including the BEH mechanism is a renormalisable quantum field theory with well controlled ultraviolet behaviour. So, individually, are its QED and QCD separate components.

In the last chapters, we have considered quantum field theory effects with running couplings and related connections to the Higgs vacuum stability and QCD confinement, as well as the role of the trace anomaly in describing hadron masses. In this chapter, we say a bit more about renormalisation in connection with the ultraviolet completion of the theory. Key issues are connections between renormalisability and gauge symmetries for theories with vector particles as well as the spacetime dimensionality, renormalisation group fixed points and quantum field theory anomalies. It is important to ensure that the quantum effects associated with vacuum polarisation are treated in calculations consistent with fundamental symmetries like gauge and Lorentz invariance when connecting with the results of experiments.

Beyond the tree approximation (Born terms) Feynman loop diagrams come with infinities that have to be controlled. For the Standard Model these ultraviolet divergences from loop diagrams can be self-consistently absorbed in a redefinition of the parameters via so-called renormalisation counterterms yielding finite renormalised quantities to be compared to experiments. The renormalisation counterterms parametrise our ignorance of deep ultraviolet physics (here including what might be beyond any scale of emergence for an emergent Standard Model). Sensible — renormalisable — theories are such that all such infinities can be absorbed in a finite number of renormalisation counterterms. The divergent counterterms and renormalised fields and interaction parameters can then be controlled and manipulated in the calculation of physical observables such as S-matrix elements.

In QCD the fields and their couplings and masses are redefined in the renormalisation procedure. The renormalised quantities (those dressed by interactions) and the corresponding bare quantities (direct from the Lagrangian and defined at the ultraviolet cut-off) are related by

$$A^a_{\text{bare } \mu} = Z_3^{1/2} A^a_{\text{ren } \mu},$$

$$\chi^a_{\text{bare } 1,2} = \tilde{Z}_3^{1/2} \chi^a_{\text{ren } 1,2}, \qquad (4.1)$$

$$\psi_{\text{bare}} = Z_2^{1/2} \psi_{\text{ren}}.$$

Here, A^a_μ denotes the gluon field, the χ^a are Fadeev–Popov ghosts and ψ represents the quark fields. Likewise, the gauge coupling g_s with $\alpha_s = g_s^2/4\pi$, the gauge parameter α (e.g. in a covariant gauge) and the quark mass m are related by

$$g_{s,\text{bare}} = Z_g g_{s,\text{ren}},$$

$$\alpha_{\text{bare}} = Z_3 \alpha_{\text{ren}}, \qquad (4.2)$$

$$m_{\text{bare}} = Z_m m_{\text{ren}}.$$

The renormalisation constants $Z_3^{1/2}$, $\tilde{Z}_3^{1/2}$, $Z_2^{1/2}$, Z_g and Z_m absorb the divergent counterterms and are calculated in [Muta (1987)]. The renormalisation of QED is discussed in [Bjorken and Drell (1965)]. The extension to the electroweak Standard Model including BEH phenomena is discussed in [Jegerlehner (1990); Kraus (1998)].

Interestingly, if a renormalisable theory includes spin $J = 1$ fields, then it is a gauge theory ['t Hooft (1980b)]. The $J = 1$ fields are gauge bosons. Gauge invariance and renormalisability then constrain the possible global symmetries of the Lagrangian. One thus finds deep connections between the internal gauge symmetries and the global symmetries with the self-consistent ultraviolet behaviour of the theory.

In $3 + 1$ dimensions one finds just logarithmic divergences in the Z_i factors for most observables. The exceptions are the renormalisation of scalar mass terms and vacuum zero-point energy contributions to the cosmological constant. Gauge boson masses are protected by gauge invariance. Fermion mass terms are protected by chiral symmetry, with counterterms proportional to the fermion masses with just logarithmic renormalisation group scale dependence. Quadratic divergences enter with the square of scalar boson mass terms and are important with the renormalisation of the Higgs boson's mass. Zero-point energies associated with vacuum energy come with quartic divergences. These quadratic and quartic divergences are discussed in detail in Chapter 10.

An important issue with renormalisation is the number of spacetime dimensions and the mass dimensionality of the coupling constants. The mass dimensionality of the momentum integrals determines the accompanying degree of divergence. The mass dimensions of fermion fields ψ, scalar bosons ϕ and the vector fields A_μ that feature in interaction terms are $[\psi] = \frac{3}{2}$, $[\phi] = 1$, $[A_\mu] = 1$. These are combined with $[m] = 1$ and $[\partial_\mu] = 1$ to give e.g. $[m\bar{\psi}\psi] = 4$ and $[\partial^\mu\phi\,\partial_\mu\phi] = 4$. The mass dimension of the Standard Model Lagrangian density is $+4$ so that the mass dimension of the action comes out as zero in natural units. With dimensionless couplings (the gauge, Yukawa and Higgs self-couplings) the theory is renormalisable. If one adds extra terms with new inverse mass dimensional couplings, then one finds extra divergences and the theory becomes non-renormalisable with bad ultraviolet behaviour signally the need for a more fundamental theory beyond the energy of the new mass scale. For example, in the early Fermi four-fermion theory of weak interactions the interaction coupling came with inverse mass squared dimension, $\frac{1}{\sqrt{2}}G_F = \frac{g^2}{8m_W^2}$, times the four-fermion vertex for the neutron $\beta-$decay $n \to pe^-\bar{\nu}_e$. In this theory the loop momentum integrals become too divergent to be renormalisable. The theory also violates perturbative unitarity. This result meant that some more precise and rigorous theory — the electroweak Standard Model with good high energy behaviour — was required at higher energies typically of size $E \approx m_W$ and above. A further example involves the effective chiral Lagrangians for low energy QCD where the dimensionful pion decay constant appears in the interaction terms and the low energy Lagrangian is not renormalisable. A consistent description of high momentum behaviour requires the full renormalisable QCD.

If the theory is formulated in extra spacetime dimensions, then the dimensionless action tells us that the gauge and Yukawa couplings would acquire a non-zero mass dimension. One would find too divergent integrals so that the theory would be non-renormalisable. QED and QCD are renormalisable in $d = 4$ spacetime dimensions, super-renormalisable (with just a finite number of divergences) in $d <$ four dimensions and non-renormalisable in $d >$ four dimensions. Detailed examples of the effect of spacetime dimensionality involving the simpler scalar ϕ^3 (renormalisable in six dimensions) and ϕ^4 theories are given in [Muta (1987)].

Beyond the renormalisation of the field operators, couplings and masses, in general one finds extra renormalisation constants for composite operators like the contributions to the energy momentum tensor involved in the trace anomaly. Conserved and partially conserved currents, with just mass terms in their divergence equations, are not renormalised. These operators come without ultraviolet divergences and thus are renormalisation scale invariant — see Section 13-5-2 of [Itzykson and Zuber (1980)].

4.1   Renormalisation Group

Renormalisation introduces some cut-off like scale when we isolate and control the divergences in momentum integrals. This renormalisation scale enters even in massless theories.[1] How the parameters of theory, that is the couplings and masses, depend on this scale is determined by the renormalisation group. We have already seen the effect of running couplings and masses in Chapters 2 and 3.

Quantities measured by experimentalists depend just on the external kinematic variables and not the details how a theoretician set up the calculation. That is, S-matrix elements depend on the kinematic variables and not the choice of renormalisation procedure including the renormalisation scale employed in a theoretical calculation, and also not on the choice of the gauge fixing procedure. In practical calculations it is common to set the renormalisation scale equal to some external momentum to simplify the calculations. For example, the fine structure constant is defined as $\alpha(m_e^2)$ in Eq. (2.7). Also, the proton structure functions in deep inelastic scattering are commonly defined with the square of the renormalisation scale μ^2 set equal to the virtuality Q^2 of the hard photon used to probe the structure of the proton in the deep inelastic experiments. The renormalisation group scale dependence of individual masses and couplings is given through β–functions and anomalous dimensions, e.g. the running couplings and masses seen in Chapters 2 and 3. The necessity of introducing a finite renormalisation scale breaks the scale invariance of the classical theory that one finds with massless fermions. The result is non-conservation of

---

[1]Renormalisation introduces a mass scale even in massless theories, e.g. from any cut-off used to define "infinite" momenta or to restore dimensional consistency when analytically continuing the dimensionality of momentum integrals with the divergences appearing in the limit of four spacetime dimensions.

the dilatation current with the trace anomaly that we saw in connection with the proton mass in Eq. (2.45).

An important property of the renormalisation group is fixed points, points in the renormalisation group evolution where the β–function is equal to zero. At renormalisation group fixed points quantum field theories theories do not depend on the variation of the renormalisation scale. At the fixed point the system behaves as a conformal field theory. (The relation between scale and conformal invariance is discussed in [Nakayama (2015)].) We have seen examples of fixed points in the running of QED, QCD and Standard Model couplings in Chapters 2 and 3. The renormalisation group can be extended beyond perturbation theory [Berges *et al.* (2002)]. Here, an interesting idea is that of asymptotic safety. This speculates that a possible non-perturbative fixed point for quantised gravitation might resolve issues with otherwise uncontrolled divergences which render a quantised version of General Relativity not perturbatively renormalisable [Weinberg (1976)]. Interestingly, it appears that all known examples of four-dimensional quantum field theories with asymptotic freedom or asymptotic safety at weak coupling involve non-abelian gauge interactions [Bond and Litim (2019)]. Quantum fluctuations of matter fields alone, with or without photons, are incapable of generating a well-defined and predictive short-distance limit at weak coupling. This result follows from the so called "a theorem" [Luty *et al.* (2013); Polchinski (1988); Shore (2017)]. Any Lorentz invariant and unitary four-dimensional quantum field theory which is under perturbative control in the deep ultraviolet or infrared asymptotes to a conformal field theory.

4.2 Symmetries and Ultraviolet Regularisation

A first step in renormalisation involves ultraviolet regularisation: choosing a mathematical procedure to define what we mean by "infinity" and to control the ultraviolet divergences. This regularisation is more than just a mathematical trick and comes with real physical input. It breaks some symmetries of the classical theory which are not necessarily all restored in the continuum [Shifman (1991)]. One has to be sure that the regularisation and final calculation preserves the fundamental ingredients of gauge invariance and Lorentz covariance with classical symmetries preserved as much as possible.[2]

---

[2]Within the emergence scenario violations of Lorentz invariance can enter with size at most $\mathcal{O}(\Lambda_{\text{ew}}^2/M^2)$ [Bjorken (2001a)].

A famous example is the axial anomaly discussed below which involves a symmetry clash from quantum corrections between local gauge invariance and chiral symmetry (a global symmetry) in triangle diagrams with the three vertices being two vector couplings γ_α and γ_β and an axial-vector vertex $\gamma_\mu \gamma_5$. Gauge invariance wins and one picks up an anomalous term in the divergence of the axial-vector current. This quantum field theory effect in the gauge invariantly renormalised current is a step beyond Noether's theorem which relates the classical version of the axial-vector current to the chiral symmetries of the Lagrangian.

Among possible regularisation procedures, the simplest idea would be to use a brute force cut-off on divergent momentum integrals. These integrals would then blow up as the cut-off is taken to infinity. However, this procedure with a fixed momentum cut-off breaks Lorentz invariance and is therefore in general not suitable for realistic calculations of many observables.

Most commonly used is the method of dimensional regularisation and minimal subtraction. One first performs a Wick rotation of the integrand and its four-momenta to Euclidean space. Dimensional regularisation then involves analytic continuation of the Euclidean spacetime dimensionality from 4 to $d = 4 - \epsilon$ dimensions so that the momentum integrals which diverge in four dimensions become finite in $4 - \epsilon$ dimensions, diverging in the limit $\epsilon \to 0$. The divergence typically appears as a pole term accompanied by the Euler constant $\gamma = 0.57721\ldots$ and $\ln 4\pi$ in the combination $\frac{1}{\epsilon} - \gamma + \ln 4\pi$. The term multiplying this divergence is subtracted out into the renormalisation constant — the so called $\overline{\text{MS}}$ scheme. Examples and subtleties with the Higgs mass squared and vacuum zero-point energies are discussed in Chapter 10. With dimensional regularisation Lorentz covariance and gauge invariance are respected, the latter with suitable treatment of Dirac γ_5 matrices in the regulator dimensions to ensure that the axial anomaly works out correctly ['t Hooft and Veltman (1972)]. The renormalisation scale enters since we need to rescale the gauge couplings $g \to g_0 \mu^{(2-d/2)}$ to maintain the dimensional consistency of physical quantities in the analytically continued momentum integrals. The connection to physical resolution is however less transparent than with a finite cut-off on the momentum integrals. With the $\overline{\text{MS}}$ procedure one typically chooses the renormalisation scale to minimise logarithms $\log m^2/\mu^2$ in perturbative calculations with the scale dependence in μ governed by renormalisation group evolution (m denotes a particle mass scale in the calculation).

Besides the $\overline{\text{MS}}$ procedure, an alternative method to define the renormalised quantities after subtracting the divergent counterterms into renormalisation constants is on-shell renormalisation. Here, the renormalised quantity becomes equal to the experimentally measured value when one takes the renormalisation scale equal to the pole mass when the corresponding particle propagates on the mass shell.

A further procedure for regularising fermion loop diagrams is the Pauli–Villars method where we control the ultraviolet divergences through replacing the fermion propagator by the propagator minus the propagator for a very heavy Pauli–Villars regulator fermion. The modified propagator returns to the original in the limit that the Pauli–Villars mass becomes infinite. This regularisation respects Poincaré invariance (translation, rotation and Lorentz boost invariance) and is a suitable regularisation in QED. Gauge invariance fails if one tries to extend this method to theories with massive gauge bosons and the BEH mechanism ['t Hooft (1971b)].

Lattice regularisation provides a successful non-perturbative procedure for controlling short distance behaviour [Creutz (1985)]. The underlying theory is formulated on a discretised Euclidean spacetime. This lattice discretisation violates the translational and rotational invariance symmetries that hold in the continuum so some care is necessary. A resulting effect is the problem of fermion doubling, that one gets too many fermion states when putting chiral fermions on the lattice. Proposed solutions involve so called Wilson fermions, Kogut–Susskind fermions or domain wall fermions; for detailed discussion see [Luscher (2002)].

In practical real world calculations it is important to choose a consistent regularisation and renormalisation procedure for all observables. Provided that the regularisation preserves the fundamental symmetries in the calculation, the final answer for observables such as S-matrix elements cannot depend on the renormalisation procedure or how a theoretician set up the calculation.

4.3 The Axial Anomaly

The fermionic axial-vector current $\bar{\psi}\gamma_\mu\gamma_5\psi$ can couple through two gauge boson intermediate states. This process involves triangle diagrams with the axial-vector current $\gamma_\mu\gamma_5$ and two vector couplings γ_α and γ_β as the three vertices — see Fig. 4.1. When we regularise the ultraviolet behavior of momenta in the triangle loop, we can preserve either current conservation

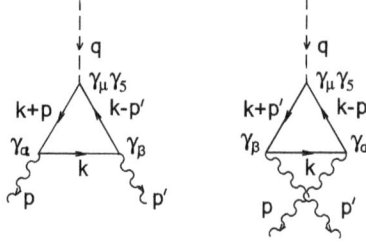

Fig. 4.1.   The diagrams, representing the vacuum expectation value of axial current in the presence of external electromagnetic field in QED, (a) the direct diagram, (b) the crossing diagram.

at the quark-gluon-vertices (necessary for gauge invariance) or partial conservation of the axial-vector current but not both simultaneously. Gauge invariance wins and induces an anomalous term in the singlet divergence equation for the axial-vector current [Adler (1969); Bell and Jackiw (1969)] which is missed in the brute force cut-off theory [Bass *et al.* (1991)].

Specialising to the QCD case, the anomaly appears in the divergence of the gauge invariantly renormalised flavour-singlet axial-vector current

$$J_{\mu 5}^{(0)} = \sum_{q=u,d,s} \bar{\psi}_q \gamma_\mu \gamma_5 \psi_q, \tag{4.3}$$

where gluonic intermediate states enter with equal coupling to u, d, s quarks. One finds the divergence equation

$$\partial^\mu J_{\mu 5}^{(0)} = \sum_{q=u,d,s} 2m_q \bar{\psi}_q i \gamma_5 \psi_q + 3 \frac{\alpha_s}{4\pi} G_{\mu\nu}^a \tilde{G}_a^{\mu\nu} \tag{4.4}$$

with the anomalous gluonic term $3\frac{\alpha_s}{4\pi} G_{\mu\nu}^a \tilde{G}_a^{\mu\nu}$ coming from the ultraviolet point-like part of the triangle loop. Here, $G_{\mu\nu}^a$ is the gluon field tensor and $\tilde{G}_{\mu\nu}^a = \frac{1}{2}\epsilon_{\mu\nu\rho\sigma} G^{a\,\rho\sigma}$ the corresponding dual tensor. The gauge invariant current can be written as the sum

$$J_{\mu 5}^{(0)} = J_{\mu 5}^{(0),\text{con}} + 2f K_\mu \tag{4.5}$$

with $\partial^\mu K_\mu = \frac{\alpha_s}{8\pi} G \cdot \tilde{G}$ being a total divergence involving the gauge dependent Chern–Simons current

$$K^\mu = \frac{g_s^2}{32\pi^2} \epsilon^{\mu\alpha\beta\gamma} A_\alpha^a \left(G_{\beta\gamma}^a - \frac{1}{3} g_s c^{abc} A_\beta^b A_\gamma^c \right) \tag{4.6}$$

and

$$\partial^\mu J^{(0),\mathrm{con}}_{\mu5} = \sum_q 2m_q \bar\psi_q i\gamma_5 \psi_q. \tag{4.7}$$

Here, $\alpha_s = g_s^2/4\pi$ and $f = 3$ the number of active light flavours. For a gluon field A_μ with gauge transformation $\mathcal{G}$, $A_\mu \to \mathcal{G}A_\mu\mathcal{G}^{-1} + \frac{i}{g_s}(\partial_\mu\mathcal{G})\mathcal{G}^{-1}$, the operator K_μ transforms as

$$K_\mu \to K_\mu + i\frac{g_s}{8\pi^2}\mathrm{Tr}\,\epsilon_{\mu\nu\alpha\beta}\partial^\alpha\big(\mathcal{G}^{-1}\partial^\nu\mathcal{G}A^\beta\big)$$

$$+ \frac{1}{24\pi^2}\mathrm{Tr}\,\epsilon_{\mu\nu\alpha\beta}\big((\mathcal{G}^{-1}\partial^\nu\mathcal{G})(\mathcal{G}^{-1}\partial^\alpha\mathcal{G})(\mathcal{G}^{-1}\partial^\beta\mathcal{G})\big), \tag{4.8}$$

where the third term on the right hand side is associated with the gauge field topology [Crewther (1978); Cronstrom and Mickelsson (1983)]. The partially conserved axial-vector current $J^{\mathrm{con}}_{\mu5}$ is not gauge invariant. The anomalous divergence equation (Eq. (4.4)) is exact as an operator equation; that is, there is no renormalisation of the anomaly coefficient $\alpha_s/4\pi$[Adler and Bardeen (1969)].

When the axial anomaly is evaluated using dimensional regularisation one uses the Dirac matrix γ_5 convention where γ_5 anti-commutes with the Dirac matrices γ^μ in the physical $3+1$ dimensions and commutes with γ^μ in the extra regulator dimensions. If we regularise the triangle diagram with a brute force cut-off on the (transverse) momentum, then the anomaly represents a contribution from beyond the cut-off [Bass *et al.* (1991)]. Non-perturbatively, quantum field theory anomalies may be understood as a collective motion of charged fermions with arbitrarily high momentum in the vacuum [Gribov (1981); Kharzeev (2011)]. Excellent reviews of anomaly physics are given in [Shifman (1991)] and [Ioffe (2006)].

Renormalisation group issues are important with $J^{(0)}_{\mu5}$. The current $J^{(0)}_{\mu5}$ is multiplicatively renormalised and renormalisation group scale dependent with an anomalous dimension that starts at two loops in perturbation theory, $\gamma(\alpha_s) = \frac{\alpha_s^2}{2\pi^2}f + \cdots$ It follows that $J^{\mathrm{con}}_{\mu5}$ is renormalisation scale invariant and the scale dependence of $J_{\mu5}$ is carried by K_μ. This is summarised in the equations

$$J^{(0)}_{\mu5} = Z_5\, J^{(0)}_{\mu5}\big|_{\mathrm{bare}}$$

$$K_\mu = K_\mu\big|_{\mathrm{bare}} + \frac{1}{2f}(Z_5 - 1)J^{(0)}_{\mu5}\big|_{\mathrm{bare}}, \tag{4.9}$$

$$J^{(0),\mathrm{con}}_{\mu5} = J^{(0),\mathrm{con}}_{\mu5}\big|_{\mathrm{bare}}$$

where Z_5 denotes the renormalisation factor for $J_{\mu5}^{(0)}$. Gauge transformations shuffle a scale invariant operator quantity between the two operators $J_{\mu5}^{(0),\mathrm{con}}$ and K_μ whilst keeping $J_{\mu5}^{(0)}$ invariant. A renormalisation scale invariant version of the gauge invariant current can be constructed: $J_{\mu5}^{(0)}/E(\alpha_s)$, where $E(\alpha_s) = e^{\int_0^{\alpha_s} d\tilde{\alpha}_s\, \gamma(\tilde{\alpha}_s)/\beta(\tilde{\alpha}_s)} = 1 - \frac{\alpha_s}{4\pi} \times \frac{24f}{(33-2f)} + \cdots$ is a renormalisation group factor that divides out the renormalisation scale dependence of $J_{\mu5}$ and which goes to one in the limit $Q^2 \to \infty$; $\beta(\alpha_s) = \mu^2 \frac{d}{d\mu^2}\alpha_s = -\alpha_s^2 \times (33-2f)/12\pi + \cdots$ is the QCD β-function. Since gluons couple equally to each flavor of quark the anomaly term cancels in the divergence equations for non-singlet currents like $J_{\mu5}^{(3)} = \bar{u}\gamma_\mu\gamma_5 u - \bar{d}\gamma_\mu\gamma_5 d$ and $J_{\mu5}^{(8)} = \bar{u}\gamma_\mu\gamma_5 u + \bar{d}\gamma_\mu\gamma_5 d - 2\bar{s}\gamma_\mu\gamma_5 s$. These operators are then partially conserved and not renormalised in QCD. They are renormalisation scale invariant in QCD.

If we also allow coupling to photons, then the flavour singlet axial-vector current satisfies the anomalous divergence equation

$$\partial^\mu J_{\mu5}^{(0)} = \sum_q 2m_q \bar{\psi}_q i\gamma_5 \psi_q + N_f \frac{\alpha_s}{4\pi} G\cdot\tilde{G} + \sum_q e_q^2 \frac{\alpha}{2\pi} F\cdot\tilde{F}. \qquad (4.10)$$

Note the relative factor of $\frac{1}{2}$ between the QED and QCD terms which comes from the different gauge group factors. Generalising the previous discussion, the axial-vector current for each flavour of quark $J_{\mu5} = \bar{\psi}\gamma_\mu\gamma_5\psi$ is now written as

$$J_{\mu5} = J_{\mu5}^{\mathrm{con}} + K_\mu + k_\mu, \qquad (4.11)$$

where again $\partial^\mu J_{\mu5}^{\mathrm{con}} = 2m_q \bar{\psi}_q i\gamma_5 \psi_q$ and K_μ is the gluonic Chern–Simons current. The QED part of the anomalous divergence is

$$\partial^\mu k_\mu = \frac{\alpha}{2\pi} F\cdot\tilde{F}, \qquad (4.12)$$

where k_μ is the gauge dependent photonic current

$$k^\mu = \frac{e^2}{16\pi^2}\epsilon^{\mu\alpha\beta\gamma}A_\alpha F_{\beta\gamma} = \frac{\alpha}{4\pi}\epsilon^{\mu\alpha\beta\gamma}A_\alpha F_{\beta\gamma}. \qquad (4.13)$$

For the isovector current $J_{\mu5}^{(3)} = \bar{u}\gamma_\mu\gamma_5 u - \bar{d}\gamma_\mu\gamma_5 d$ the gluonic current K_μ cancels between up and down flavoured quarks. The anomalous divergence then reads

$$\partial^\mu J_{\mu5}^{(3)} = (2m_q\bar{\psi}_q i\gamma_5 \psi_q)_{u-d} + (e_u^2 - e_d^2)\frac{\alpha}{2\pi}F\cdot\tilde{F}. \qquad (4.14)$$

This QED anomaly plays an essential role in the $\pi^0 \to 2\gamma$ decay, which would otherwise vanish in the chiral limit with massless quarks and $m_\pi = 0$ [Adler (1969); Bell and Jackiw (1969)]. Chiral corrections from finite quark masses give an about 4.5% extra contribution to the π^0 decay rate beyond the anomaly term [Bernstein and Holstein (2013)].

4.3.1 *Anomaly cancellation in the Standard Model*

In the Standard Model, the axial anomaly plays an essential role in grouping the fermions into families [Bouchiat *et al.* (1972); Gross and Jackiw (1972)]. The Standard Model includes triangle loop diagrams where each of the three vertices couple to gauge bosons, including to SU(2) weak bosons with just left-handed coupling to fermions. Gauge invariance and renormalisability tell us that there should be no net anomaly contribution in these loops after we sum over all fermions propagating in the loop, e.g. the WWγ triangle loop. This constraint groups the fermions into families with combinations of charges perfectly aligned to cancel any local chiral anomalies. For each family the sum over the fermions' electric charges should vanish after we also take into account the three colours for propagating quark species to cancel any local chiral anomalies, viz.

$$\sum_i Q_i = 0, \tag{4.15}$$

for example,

$$\{e_e + 0 + N_c(e_u + e_d)\} = \left\{-1 + 0 + 3\left(\frac{2}{3} - \frac{1}{3}\right)\right\} = 0. \tag{4.16}$$

4.3.2 *Definitions of chirality and baryon number*

The axial anomaly gives us multiple forms of the axial-vector current: $J_{\mu 5}$ and $J_{\mu 5}^{\text{con}}$. One consequence is that the definitions of both chirality and baryon number in the Standard Model become quite subtle. In the singlet channel there is no classical Noether axial-vector current corresponding to chiral symmetry once the anomaly becomes important [Adler (1970); Bass (2004); Crewther (1978)]. As observables, chirality and baryon number should be gauge invariant as well as independent of the renormalisation scale. Further, the gauge bosons should not carry chirality and/or baryon number.

First consider QCD chirality. We choose the $A_0 = 0$ gauge and define two operator charges [Crewther (1978)]

$$X(t) = \int d^3z\, J_{05}(z), \quad Q_5 = \int d^3z\, J_{05}^{\text{con}}(z). \qquad (4.17)$$

Because conserved currents are not renormalised it follows that Q_5 is a time independent operator. The charge $X(t)$ is manifestly gauge invariant whereas Q_5 is invariant only under "small" gauge transformations; the charge Q_5 transforms as

$$Q_5 \to Q_5 - 2f\, n, \qquad (4.18)$$

where n is the winding number associated with the gauge transformation $\mathcal{G}$. Although Q_5 is gauge dependent we can define a gauge invariant chirality q_5 for a given operator $\mathcal{O}$ through the gauge invariant eigenvalues of the equal-time commutator

$$[Q_5,\ \mathcal{O}]_- = -q_5\, \mathcal{O}. \qquad (4.19)$$

The gauge invariance of q_5 follows since this commutator appears in gauge invariant Ward Identities despite the gauge dependence of Q_5. The time derivative of spatial components of the gluon field have zero chirality q_5 but non-zero X charge:

$$[Q_5,\ \partial_0 A_i]_- = 0 \qquad (4.20)$$

and

$$\lim_{t'\to t}\left[X(t'),\ \partial_0 A_i(\vec{x}, t)\right]_- = \frac{i f g_s^2}{4\pi^2}\tilde{G}_{0i} + O(g_s^4 \ln|t' - t|). \qquad (4.21)$$

Equation (4.20) follows from the non-renormalisation of the partially conserved current $J_{\mu5}^{\text{con}}$. Equation (4.21) follows from the implicit A_μ dependence of the (anomalous) gauge invariant current $J_{\mu5}$. The higher-order terms $g_s^4 \ln|t' - t|$ are caused by wavefunction renormalisation of $J_{\mu5}$ [Crewther (1978)]. The analogous situation in QED is discussed in [Adler (1970); Adler and Boulware (1969); Jackiw and Johnson (1969)].

Thus, defining chirality through the gauge invariant commutators involving $J_{\mu5}^{\text{con}}$ which appear in anomalous Ward identities provides an optimal solution. The eigenvalues are gauge and renormalisation scale invariant. Also, the gauge fields carry no net chirality. With this definition through $J_{\mu5}^{\text{con}}$ and Q_5, the chiralities of the delocalised quark-antiquark pairs found in the QCD topological θ vacuum — Section 2.2.2 — cancel against a gluonic term measuring the topological winding number to give zero net axial-charge in each of the θ-vacuum substates $|m\rangle$. Similarly, in the light-front gauge $A_+ = 0$, the forward proton matrix element of J^{con}_{+5} corresponds to the renormalisation scale invariant $\Delta q_{\text{partons}}$ contribution in polarised deep inelastic scattering (Eq. (2.51)). The corresponding "quark spins" satisfy the SU(2) Pauli spin algebra without an extra renormalisation group factor appearing on one side of the spin commutator relations. Otherwise, with "spin" defined via $J_{\mu 5}$, if the quark spin operators satisfy SU(2) commutator relations at one scale then they do not satisfy the SU(2) relations at a different scale because of the two-loop anomalous dimension of $J_{\mu 5}$ [Bass and Thomas (1993)]. The polarised gluon contribution $-\frac{\alpha_s}{2\pi}\Delta g$ corresponds to the "+ component" of the forward matrix of the gluonic Chern–Simons current K_+ evaluated in light-front gauge $A_+ = 0$ modulo possible zero-mode topology contributions associated with the $\mathcal{C}_\infty$ term in Eq. (2.51).

4.3.3 Baryon number and the axial anomaly

The axial anomaly is also important to how we define baryon and lepton number in the Standard Model. The vector baryon/lepton current can be written as the sum of left and right handed currents,

$$J_\mu = \bar{\Psi}\gamma_\mu\Psi = \bar{\Psi}\gamma_\mu\frac{1}{2}(1 - \gamma_5)\Psi + \bar{\Psi}\gamma_\mu\frac{1}{2}(1 + \gamma_5)\Psi. \qquad (4.22)$$

In QED and QCD this current is conserved with just parity conserving vector couplings. In the Standard Model just the left handed fermions couple to the SU(2) electroweak gauge bosons. This means that the baryon/lepton current is sensitive to the axial anomaly ['t Hooft (1976b)]. One finds the anomalous divergence equation

$$\partial^\mu J_\mu = n_f\left(-\partial^\mu \mathcal{K}_\mu + \partial^\mu k_\mu\right), \qquad (4.23)$$

where $\mathcal{K}_\mu$ and k_μ are the SU(2) electroweak and U(1) hypercharge anomaly currents

$$\mathcal{K}_\mu = \frac{g^2}{16\pi^2}\epsilon_{\mu\nu\rho\sigma}\left[W_a^\nu\left(\partial^\rho W_a^\sigma - \frac{1}{3}gf_{abc}W_b^\rho W_c^\sigma\right)\right] \qquad (4.24)$$

and

$$k_\mu = \frac{g'^2}{16\pi^2}\epsilon_{\mu\nu\rho\sigma}B^\nu\partial^\rho B^\sigma. \qquad (4.25)$$

Here, W_μ and B_μ denote the SU(2) and U(1) gauge fields, with the corresponding anomalous currents satisfying $\partial^\mu \mathcal{K}_\mu = \frac{g^2}{32\pi^2} W_{\mu\nu} \tilde{W}^{\mu\nu}$ and $\partial^\mu k_\mu = \frac{g'^2}{32\pi^2} F_{\mu\nu} \tilde{F}^{\mu\nu}$. One finds a conserved but gauge dependent current

$$J_\mu^{\mathrm{con}} = J_\mu - n_f(-\mathcal{K}_\mu + k_\mu) \tag{4.26}$$

with divergence equation

$$\partial^\mu J_\mu^{\mathrm{con}} = 0. \tag{4.27}$$

The current J_μ^{con} is SU(2) and U(1) gauge dependent through the gauge dependence of $\mathcal{K}_\mu$ and k_μ.

The gauge invariantly renormalised current J_μ is scale dependent with the two-loop anomalous dimension induced by the axial anomalies associated with $\mathcal{K}_\mu$ and k_μ. Just the J_μ^{con} is renormalisation scale invariant. Similar to the discussion with chirality, the commutators of the symmetry current charge give a renormalisation scale invariant and gauge invariant (under both large and small gauge transformations) definition of baryon/lepton number with vanishing value for the W and Z boson gauge fields and their time derivatives [Bass (2004)].

Chapter 5

Emergent Gauge Symmetries and Particle Physics

Today we have the situation where the Standard Model is very successful in describing the results of our present experiments. There is no evidence (so far) for new particles or interactions in the domain of the experiments. Yet we know that new physics is required with open puzzles waiting to be explained. When taken alone the Standard Model has nice ultraviolet behaviour including a Higgs vacuum that is (almost) stable up to the Planck scale. How high in energy might it work before the entrance of new particles and/or interactions? Where at a deeper level do the gauge symmetries of particle physics come from? Might the Standard Model be more special than previously thought? If so, how might we resolve the open questions in particle physics?

Here, we consider the possibility that the Standard Model might work up to very large scales where the Higgs self-coupling λ continues to remain positive under renormalisation group evolution. In this chapter, we develop the idea of an emergent Standard Model. The key idea involves the Standard Model including its particles and gauge symmetries being "born" in some phase transition deep in the ultraviolet. The Standard Model particles including the Higgs and gauge bosons would then be the long-range, collective excitations of the critical system that resides above the phase transition.

Emergent gauge symmetries are observed in quantum condensed matter systems in connection with topological phases where one finds new gauged $J = 1$ quasiparticles among the low energy excitations. These excitations correspond to extra emergent gauge symmetries that are dynamically generated in a phase transition and are extra to the fundamental QED dynamics provided by photon exchanges. The phase transitions that produce the emergent gauge symmetries come without local order parameters, thus differing from usual Ginzburg–Landau type phase transitions. These so called quantum topological phase transitions are associated with topological order and long-range quantum entanglement. Local gauge symmetries act just on internal degrees of freedom instead of relating physical states so we do not expect any local order parameter associated with their appearance. Examples of emergent gauge symmetries in condensed matter physics include in the Fermi–Hubbard model, high temperature superconductors, spin ice, the quantum Hall effect and the A-phase of superfluid ^{3}He as well as string-net phenomena. These condensed matter systems are discussed in Chapter 6. In general, there are well known connections between quantum field theories and statistical Ising-like systems. This theory with emphasis on Ginzburg–Landau type phase transitions is discussed in detail in the monographs [Zinn-Justin (1989, 2007)]. The extension to topological order and phase transitions is a step beyond this.

Besides in these phase transitions, emergent gauge symmetries can also arise at an infrared fixed point in the renormalisation group with decoupling of gauge symmetry violating degrees of freedom [Wetterich (2017)] and in connection with any spontaneously breaking of Lorentz symmetry [Bjorken (2001a, 1963); Chkareuli *et al.* (2001)].

5.1   An Emergent Standard Model

How might emergent gauge symmetries work in particle physics and might the Standard Model be emergent? Gauge symmetries can be emergent below a scale deep in the ultraviolet.

Consider a statistical system near its critical point. The long range asymptote is a renormalisable quantum field theory with properties described by the renormalisation group [Peskin and Schroeder (1995); Wilson and Kogut (1974)]. If the spectrum includes $J = 1$ excitations among the degrees of freedom in the low energy phase, then it is a gauge theory. Gauge symmetries would then be an emergent property of the low energy phase and "dissolve" in some topological like phase transition deep in the ultraviolet [Bass (2021); Bjorken (2001a); Forster *et al.* (1980); Jegerlehner (1978, 1998, 2014c); 't Hooft (2007)]. The quarks and leptons as well as the gauge bosons and Higgs boson would then be the stable collective long-range excitations of some (unknown) more primordial degrees of freedom that exist above the scale of emergence. The vacuum of the low energy phase should be stable below the scale of emergence.

Suppose the Standard Model works like this. Then it behaves like an effective theory with the renormalisable theory at mass dimension four, $D = 4$, supplemented by a tower of non-renormalisable higher dimensional operators suppressed by factors of the large scale of emergence. The global symmetries of the Standard Model at $D = 4$ are constrained by gauge invariance and renormalisability. The higher dimensional operators are less constrained and may exhibit extra global symmetry breaking [Bass (2020, 2021); Jegerlehner (2014c); Weinberg (2018); Witten (2018)]. Lepton number violation and tiny Majorana neutrino masses may enter at $D = 5$, suppressed by a single power of the scale of emergence via the so called Weinberg operator [Weinberg (1979)]. One finds $m_\nu \sim \Lambda_{\text{ew}}^2/M$ with Λ_{ew} the electroweak scale and M the scale of emergence. Possible proton decays might occur at $D = 6$ suppressed by two powers of M [Weinberg (1979); Wilczek and Zee (1979)]. New CP violation, needed for baryogenesis, might occur in Majorana neutrino phases at $D = 5$ as well as in new $D = 6+$ operators [Grzadkowski *et al.* (2010)]. If we identify the neutrinos of particle physics with the Majorana neutrinos predicted here with $m_\nu \sim \Lambda_{\text{ew}}^2/M$, then we expect $M \sim 10^{16}$ GeV.

If one increases the energy much above the electroweak scale, then the physics becomes increasingly symmetric with energies $E \gg \Lambda_{\text{ew}}$ until we come within about 0.1% or so of the scale of emergence. Then new global symmetry violations in higher dimensional operators become important so the physics becomes increasingly chaotic until one goes through the phase transition associated with the scale of emergence, with the physics above this scale then described by new degrees of freedom and perhaps new physical laws. This scenario contrasts with unification models which involve maximum symmetry in the extreme ultraviolet. The effect of the higher dimensional operators might have been especially active in the very early Universe with energies $E \sim M$ and be manifest in observables characterised by tiny masses and couplings in the Universe that we see today.[1]

---

[1] In the formal language of renormalisation group the operators with mass dimension less than four are called relevant operators, dimension four operators are called marginal operators and higher dimensional operators are called irrelevant operators. The few relevant and marginal operators can be invariant under a wider range of field transformations than a generic irrelevant operator would be. The effects of irrelevant operators are strongly suppressed at low energies (suppressed by powers of the large emergence scale) making it appear that the theory has a larger symmetry group. Symmetry can be emergent in the low energy theory even if it is not present in the underlying microscopic theory, e.g. associated with an infrared fixed point.

Emergence works fundamentally different to unification models. With unification one expects the running gauge couplings of the Standard Model to meet in the ultraviolet corresponding to the electroweak and QCD interactions unifying within some larger gauge group. They do come close (see Fig. 3.1) but without exactly crossing. Crossing could be achieved with the addition of supersymmetry [Wess and Zumino (1974)], SUSY, at TeV energies [Ross and Roberts (1992)]. Supersymmetry models postulate a new symmetry between bosons and fermions and can also provide a dark matter candidate particle. Although the simplest SUSY models would have preferred a Higgs boson mass close to the measured value, the absence of any sign of new SUSY particles in the LHC experiments means that these models are now strongly constrained [Altarelli (2013b, 2014); Pokorski (2016); Ross (2014)]. As a general comment on theories of possible new physics beyond the Standard Model, any new symmetries that might arise in the ultraviolet connecting interactions at $D = 4$ must be strongly broken so that they are not seen at the energies of current experiments. This implies a trade off: the extra symmetry that might exist at higher energies also comes with a (perhaps large) number of new parameters needing extra explanation.

When discussing the renormalisation group and phase transitions an important concept is that of universality. The physics details above the characteristic energy scale for a phase transition can be quite different but these differences may be washed out when going through the phase transition and lead to very similar low energy behaviour as seen at large distances in the low energy theory. For an extended discussion see the books [Palacios (2022); Zinn-Justin (2007)]. High energy systems which flow towards the same fixed point are said to be in the same universality class. Initial values of the coupling constants do not determine the critical behaviour. Only the couplings to a few marginal and relevant operators do. With an emergent Standard Model the theoretical challenge is to identify the universality class of theories which have the Standard Model (plus any new particle interactions waiting to be discovered) as their long range asymptote [Jegerlehner (2014c)].

There is also the interesting issue of the critical dimension associated with a phase transition. With a statistical Ising system near its critical point, the Ginzburg criterion tells us that fluctuations become important for spacetime dimensions of four or less. This coincides with the dimensionality of spacetime. With four spacetime dimensions one finds an interacting quantum field theory — ϕ^4 in Euclidean space. Analyticity allows us to use the Wick rotation meaning that the Euclidean quantum field theory is mathematically equivalent to the theory in Minkowski space. With $d > 4$ spacetime dimensions the physics reduces to a free field theory with long range modes decoupled. For $d < 4$, the theory exhibits scaling behaviour described by the Wilson–Fisher fixed point. The case of $d = 4$ dimensions is at the border where the ϕ^4 theory long-range tail is precisely renormalisable. The dimensional analysis predictions are corrected but only by logarithms.

With emergent gauge symmetries the topological structure of the Yang–Mills vacuum will also be present with gauge invariance taken to include invariance under both "small" and "large" gauge transformations. The clusterisation principle discussed in Chapter 2 tells us that invariance under large gauge transformations is essential [Shifman (1991)]. When it is coupled to gravitation an emergent Standard Model gives a simple explanation of the cosmological constant, see Chapter 9. One also finds interesting applications to the Higgs mass hierarchy puzzle (see Chapter 10) and constraints on possible dark matter scenarios as well as hints on possible applications to baryogenesis, see Chapter 11.

5.1.1 *Renormalisation group decoupling and spontaneous breaking of Lorentz symmetry*

In addition to topological-like phase transitions, emergent gauge symmetries can also appear through decoupling of gauge violating terms in the renormalisation group at an infrared fixed point [Forster *et al.* (1980); Wetterich (2017)] and in connection with possible spontaneous breaking of Lorentz symmetry, SBLS [Bjorken (2001a, 1963, 2010); Chkareuli *et al.* (2001)].

In the former case, the coefficient of any local gauge symmetry violating terms blows up at the fixed point, in contrast to the restoration of global symmetries where the coefficient of any symmetry violating term goes to zero at the fixed point. With global symmetries, enhanced symmetry is associated with a fixed point in the renormalisation group flow if the flow equation is compatible with the symmetry. A macroscopic global symmetry can be generated if the fixed point is infrared stable. With local gauge symmetries the renormalisation group flow should eliminate degrees of freedom which then no longer belong to the spectrum of physical excitations [Wetterich (2017)]. This may be achieved by generating large mass like terms for the extra degrees of freedom which decouple leaving a residual local gauge symmetry for the remaining light degrees of freedom.

The gauge fixing term might be generated in the renormalisation group flow at an attractive infrared fixed point even if not present in the microscopic description.

In addition to gauge symmetries, Lorentz invariance can also be emergent in the infrared. Nielsen and collaborators considered the effect of adding Lorentz violating terms to non-covariant generalisations of QED and Yang–Mills theory and found that they vanish in the infrared through renormalisation group evolution, e.g. with Lorentz invariance emerging at an infrared fixed point [Chadha and Nielsen (1983); Nielsen and Ninomiya (1978)].

Emergent gauge symmetries can also be linked with spontaneous breaking of Lorentz symmetry. In early work Bjorken considered the photon as a Nambu–Goldstone boson associated with spontaneous breaking of Lorentz symmetry deep in the ultraviolet [Bjorken (2001a, 1963)]. In Bjorken's picture SBLS is taken with scale M close to the scale of ultraviolet completion. The net vector field A_μ is taken as $A_\mu = a_\mu + M n_\mu$ where a_μ is the physical photon field and n_μ is a unit timelike vector characterising the SBLS. Lorentz invariance works everywhere it has so far been tested in experiments from low energy precision measurements to the highest energy cosmic rays [Shore (2005)]. Any violations will be tiny. Motivated by the experimental constraints on Lorentz invariance, possible violations were phenomenologically conjectured to be of order $\delta(\mathrm{LV}) \sim \mu_{\mathrm{vac}}/M$ where μ_{vac} is the cosmological constant scale taken as $\mu_{\mathrm{vac}} \sim \Lambda_{\mathrm{ew}}^2/M$ with Λ_{ew} the scale of electroweak symmetry breaking [Bjorken (2001a, 2001b)]. With SBLS the preferred reference frame in this picture is naturally identified with the frame where the Cosmic Microwave Background is locally at rest. Interestingly, the cosmological constant scale here $\mu_{\mathrm{vac}} \sim \Lambda_{\mathrm{ew}}^2/M$ also follows from detailed emergence arguments applied to the symmetries of the metric for a "pure vacuum" Universe, see Chapter 9 and [Bass (2023); Bass and Krzysiak (2020b)].

With SBLS, non-observability of any Lorentz violating terms at $D = 4$ corresponds to gauge symmetries for vector fields like the photon. In an alternative approach, Chkareuli, Froggart and Nielsen considered possible SBLS together with an additional requirement of non-observability of all Lorentz non-invariant SBLS terms [Chkareuli *et al.* (2001)]. Here one again considers the form $A_\mu = a_\mu + \mathcal{M} n_\mu$ in the general Lagrangian density restricted to terms of mass dimension four or less. The contribution $\mathcal{M} n_\mu$ represents a classical background field appearing when the vector field A_μ

develops a vacuum expectation value; a_μ is the physical photon field. For QED this involves the extra terms

$$\frac{1}{2}M_A^2 A_\mu A^\mu + \frac{f}{4}A_\mu A^\mu \cdot A_\nu A^\nu, \qquad (5.1)$$

where the mass term M_A and interaction coupling f are to be fixed. Requiring non-observability of terms involving n_μ, so that the net term involving n_μ vanishes, corresponds to requiring $\mathcal{M}^2 = f = 0$ for the coefficients of the possible gauge symmetry violating terms proportional to the square of the vector field. Otherwise one has to impose an additional gauge fixing term $n.a = 0$ which is inconsistent with the Lorentz gauge $\partial^\mu a_\mu = 0$ taken as already imposed on the vector field a_μ. The vector fields become the source for the gauge symmetry with the SBLS converted into the gauge degrees of freedom of the massless vector fields. This scenario compares with the usual picture where gauge invariance is the source for the vector gauge fields. This argument can be generalised to non-abelian gauge fields.

Other ideas discussing emergent Lorentz and gauge invariance are discussed in [Gomes (2016)] and [Barceló *et al.* (2016, 2021)].

Besides discussion of gauge symmetries and Lorentz invariance, quantum spin-statistics can emerge from classical models as well as models with just bosonic degrees of freedom. Fermions can be obtained from probablistic cellular automata models [Wetterich (2021, 2022)] and generalised Ising models with classical degrees of freedom [Wetterich (2010a)]. Also, in condensed matter physics inspired string-net models of QED (see Chapter 6), starting from string connected excitations of a bosonic lattice one obtains a model unification of gauge interactions and Fermi–Dirac statistics [Levin and Wen (2005a)]. The strings can themselves condense and fermions correspond to the ends of string excitations above this condensate. Quantised fermions can also arise as topological solitons — so called Skyrmions — in theories of bosons [Skyrme (1962)].

5.1.2 *Composite gauge boson models and the LHC*

Besides ideas on the possible emergence of Standard Model gauge symmetries below a scale deep in the ultraviolet, about 10^{16} GeV or more, ideas have also been considered where gauge bosons and perhaps also the quarks and leptons are constructed as composites of some more primordial fermions on a distance scale large compared to 10^{16} GeV, see e.g. [Amati *et al.* (1981); Fritzsch (2012); Fritzsch and Mandelbaum (1981); Terazawa *et al.* (1977)]. These models tend to predict new states in the LHC energy range that so far have not been seen in the experiments.

5.2 The Weinberg–Witten Theorem and Emergent Gauge Bosons

The Weinberg–Witten theorem is sometimes quoted as a strong constraint on ideas about possible emergent or composite (massless) gauge bosons [Weinberg and Witten (1980)]. However, it does come with loopholes with exemptions for photons in QED, gluons in QCD, W and Z bosons and possible gravitons [Loebbert (2008)].

The theorem comes in two parts and states:

(1) A theory that allows the construction of a Lorentz-covariant conserved four-vector current J^μ cannot contain massless fields of spin $j > \frac{1}{2}$ with nonvanishing values of the conserved charge $\int J^0 d^3x$.
(2) A theory that allows the construction of a conserved Lorentz-covariant energy-momentum tensor $\theta^{\mu\nu}$ for which $\int \theta^{0\nu} d^3x$ is the energy-momentum four-vector cannot contain massless particles of spin $j > 1$.

This theorem is evaded by the gauge bosons of Standard Model interactions. For QED the photon does not carry the charge of the electromagnetic current. Weak interactions are mediated by massive W and Z bosons with masses coming from the BEH mechanism. The gauge boson masses get them around the theorem. In QCD, the current either contains just quark degrees of freedom, so that the gluon is not charged under it, or is gauge dependent and therefore not covariant. There is also the issue of confinement that free massless gluons do not exist in Nature. Likewise, with possible gravitons associated with any quantised version of General Relativity the gravitons are not charged under a Lorentz covariant current.

5.3 The Standard Model in a Low-Energy Expansion

An emergent Standard Model comes with a tower of higher dimensional operators with associated new global symmetry violations suppressed by powers of the scale of emergence. If we regard the Standard Model as an effective theory here, then the characteristic energy becomes the scale of emergence.

New physics can be associated with the higher dimensional operators [Jegerlehner (2014c); Weinberg (2018)]. These operators might play an essential role in understanding neutrino masses as well as new sources of CP violation beyond the minimal Standard Model with new CP violating operators as well as CP phases with Majorana neutrinos. Proton decays and new pseudoscalar axion particles are also possible. These phenomena would enter suppressed by powers of the large scale of emergence. Possible applications to the cosmological constant and dark matter are discussed in Chapters 9 and 10. Fermion family structure comes for free along with anomaly cancellation required by gauge invariance and renormal- isablity in the emergent gauge system. Any unitarity violating terms like non-renormalisable higher dimensional operators will come suppressed by powers of the large scale of emergence, the limit of the effective theory.

The Standard Model effective theory discussed here differs from usual discussions in particle theory where one supposes that new interactions at $D = 4$ might enter at energies above the range of present experiments. The effect of these interactions is "integrated out" when comparing with present experiments. This involves redefinition of the low energy parameters to include the effect of higher order corrections from possible new heavy degrees of freedom. "New physics" is parametrised by higher dimensional terms with the mass scale representing the scale of the new physics. New heavy particle degrees of freedom would only be liberated when one goes through their production thresholds [Isidori *et al.* (2024)]. As a classic example, before discovery of the W and Z bosons parity violating weak interactions were described using Fermi's four-fermion interaction with coupling

$$\frac{1}{\sqrt{2}} G_F = \frac{g^2}{8m_W^2}. \tag{5.2}$$

The need for a theory with consistent high energy behaviour signaled the need for new physics at a deeper level — the electroweak Standard Model with its W and Z gauge bosons. From colliders the LHC data (so far) reveal no evidence for higher dimensional correlations divided by powers of a large mass scale below the few TeV range [Ellis *et al.* (2021); Slade (2019)], thus constraining the heavy mass suppression factor to be above the few TeV scale. In our emergence scenario the Standard Model is taken to work at mass dimension $D = 4$ up to the large scale of emergence. The higher dimensional terms describe new physics from the phase transition which produces the Standard Model but unconstrained by requirements of renormalisability at $D = 4$.

Within this discussion there is a usual assumption of exact gauge invariance in each term in the higher dimensional operator tower. Possible gauge symmetry violations in higher dimensional operators are discussed in [Bjorken (2001a)]. Corrections to exact gauge invariance in models with synthetic gauge fields and quantum simulations are discussed in [Halimeh and Hauke (2022)].

5.3.1   *Neutrino masses*

Neutrinos are the less explored part of the Standard Model.

In the Standard Model one expects three flavours of neutrinos, one corresponding to each of the three charged leptons. Standard Model fits to LEP and SLC data give 2.996 ± 0.007 types of light neutrinos [Janot and Jadach (2020); Schael *et al.* (2006)] whereas direct measurements of the invisible Z width give 2.92 ± 0.05 types of neutrinos [Navas *et al.* (2024)]. In parallel, the number of light neutrino species can be extracted from cosmology data with 2.99 ± 0.17 found from measurements of the Cosmic Microwave Background, CMB, and baryon acoustic oscillations, BAO [Aghanim *et al.* (2020)]. Cosmology constraints on neutrino properties are reviewed in [Lesgourgues and Verde (2024)].

In the minimal Standard Model neutrinos come with just left-handed chirality, zero mass and interact only through their couplings to the massive W and Z weak gauge bosons. But this picture has proven too simple. Neutrino oscillation experiments, where neutrinos created with a particular flavour (corresponding to the electron, muon or tau) are later measured to have a different flavour, point to the existence of tiny neutrino masses m_ν [Balantekin and Kayser (2018)]. These tiny neutrino masses and neutrino flavour mixing are deduced from solar, atmospheric and reactor neutrino disappearance experiments as well as from accelerator based appearance and disappearance experiments. Assuming that there are three species of neutrinos, the neutrino oscillation data constrains the largest mass squared difference to be $\approx 2 \times 10^{-3}$ eV^2 with the smaller one as $(7.53 \pm 0.18) \times 10^{-5}$ eV^2. The lightest neutrino mass is expected to be about 10^{-8} times the value of the electron mass. Neutrinos are considered with two possible mass orderings: normal ordering with $m_1 < m_2 < m_3$ and inverted ordering with $m_3 < m_1 < m_2$. Experimentally, the data show a hierarchy between the mass splittings, $\Delta m_{21}^2 \ll |\Delta m_{32}^2| \simeq |\Delta m_{31}^2|$ ith $\Delta m_{ij}^2 \equiv m_i^2 - m_j^2$. One finds [Navas *et al.* (2024)]

$$\Delta m_{21}^2 = (7.53 \pm 0.18) \times 10^{-5} \text{ eV}^2,$$

$$\Delta m_{32}^2 = (2.445 \pm 0.028) \times 10^{-3} \text{ eV}^2 \qquad \text{(normal ordering)}, \qquad (5.3)$$

$$\Delta m_{32}^2 = -(2.529 \pm 0.029) \times 10^{-3} \text{ eV}^2 \qquad \text{(inverted ordering)}.$$

Mass measurements for individual neutrino flavours remain a major experimental challenge. The present most precise direct measurement of neutrino masses comes from the KATRIN tritium β-decay experiment, $m_{\bar{\nu}_e} < 0.45$ eV [Aker *et al.* (2024)].

Complementary constraints come from cosmology. Cosmology gives a bound on the neutrino mass sum with the Particle Data Group combined value [Navas *et al.* (2024)]

$$\sum m_\nu < 0.12 \text{ eV} \qquad (5.4)$$

based on CMB, gravitational lensing and BAO data and in the context of the usual ΛCDM model of cosmology. Measurements of individual observables give bounds typically about 0.5 eV.

Based on the neutrino oscillation data one can deduced the neutrino mixing angles. The mixing angles for normal ordering come out to be [Navas *et al.* (2024)]

$$\sin^2 \theta_{12} = 0.307 \pm 0.013, \quad \sin^2 \theta_{13} = 0.0219 \pm 0.0007,$$

$$\sin^2 \theta_{23} = 0.558^{+0.015}_{-0.021}. \qquad (5.5)$$

One finds bigger mixings with neutrinos than with quarks with terms more similar in size than separated by an order of magnitude like what happens in the quark CKM matrix [Fritzsch and Xing (1998)]. Some extra source of CP violation beyond the Standard Model is needed to understand the observed matter-antimatter asymmetry in the Universe. There are hints for possible CP violation in the neutrino sector though CP symmetry is still allowed at the level of 1–2 standard deviations. Oscillation data suggest a CP violating phase $\delta = 1.19 \pm 0.22 \times \pi$ where π corresponds to no CP violation [Navas *et al.* (2024)].

Because neutrinos are weakly interacting, precision measurements of their properties are very challenging. An open question is whether neutrinos are Dirac or Majorana fermions, with Majorana particles being their own antiparticles. Majorana neutrinos have the property that each mass eigenstate with a given helicity coincides with its own antiparticle with the same helicity. A related puzzle is the origin of the tiny neutrino masses, these being very much smaller than the masses of the charged leptons and quarks. For Majorana fermions, neutrinos and antineutrinos can be identified meaning that lepton number is not conserved and not a good quantum number.

If neutrinos are Majorana particles, then their masses can be linked to lepton number violation and the dimension five Weinberg operator [Weinberg (1979)],

$$L_5 = \frac{(\Phi L)_i^T \lambda_{ij} (\Phi L)_j}{M} \qquad (5.6)$$

suppressed by a single power of a large mass scale M. The Weinberg operator involves just the Standard Model lepton fields and the BEH Higgs doublet field. In Eq. (5.6), Φ is the Higgs doublet, the L_i denote the SU(2) left-handed lepton doublets defined in Eq. (2.55) and λ_{ij} is a matrix in flavour space. This yields the neutrino masses

$$m_\nu \sim \Lambda_{ew}^2/M, \qquad (5.7)$$

where Λ_{ew} is the electroweak scale 246 GeV and M is a large mass scale. Comparing thie exxpression with expectations for the lightest neutrino mass gives the mass scale $M \sim 10^{16}$ GeV [Altarelli (2005)]. Neutrino mass measurements are thus exploring very high energy scales. The Majorana mass term $\nu_L^T \nu_L$ that appears through Eq. (5.6) violates lepton number conservation by two units, e.g. through the process $\nu \to \bar{\nu}$. The expression in Eq. (5.7) for light Majorana neutrino masses is also obtained in so called see-saw mechanism models [Gell-Mann *et al.* (1979); Glashow (1980); Minkowski (1977); Mohapatra and Senjanovic (1980); Yanagida (1979)]. These models comes with additional very heavy right handed Majorana neutrinos with mass $\approx M$. For further discussion, see [Balantekin and Kayser (2018)] and the lectures [Altarelli (2013a)].

If neutrinos are Dirac particles then one option is that there is an extra type of neutrino: sterile right-handed neutrinos without direct coupling to Standard Model gauge bosons. This scenario would require some new mechanism to explain the neutrino tiny masses compared to charged leptons and quarks, viz. why these Yukawa couplings are so much suppressed relative to the charged lepton couplings with right-handed neutrinos not participating in electroweak interactions.

A signature of Majorana neutrinos would be lepton number violation in the neutrinoless double β−decay of nuclei involving the decay of two neutrons to two protons plus electrons with no neutrinos emitted, $(A, Z) \rightarrow (A, Z+2) + e^- + e^-$, where A is the atomic mass number and Z is the atomic number. Neutrinoless double β− decay would involve two charged W gauge bosons coupled to neutrons in the nucleus. One neutron decays into an electron and electron-antineutrino. The second absorbs a neutrino to create the second electron in the final state. This only works if the neutrino and antineutrino are identical, that is, that neutrinos are their own antiparticles. Planned experiments using current technologies will reach the precision $m_{\beta\beta} < 15$ meV limit, with $m_{\beta\beta}$ the modulus of a linear combination of neutrino masses [Dolinski *et al.* (2019)]. Present experimental constraints put $m_{\beta\beta}$ today less than about 0.1 eV. An outlook to future more precise measurements is given in [Baudis (2023)]. Experimental precision could be pushed further with new quantum sensing technologies [Bass and Doser (2024)].

In general, Majorana neutrinos come with two extra CP mixing angles in addition to the CP angle associated with mixing of Dirac fermions. On the basis of present measurements one cannot say anything about the possible size of these extra angles.

5.3.2   *Proton decays and baryon number violation*

In addition to lepton number violation, there is also the possibility of baryon number violation which can arise through the four-fermion operator combinations

$$B_6 = \frac{1}{M^2} QQQL \tag{5.8}$$

which enter with $1/M^2$ suppression and with Lorentz structure described in [Weinberg (1979); Wilczek and Zee (1979)]. Here, L and Q are the lepton and quark doublets defined in Eq. (2.55). This interaction leads to two body decays like $p \rightarrow l^+ \pi^0$ with rate $\Gamma \propto M_{\text{P}}^5 / M^4$ where M_{P} is the proton mass. Present experimental sensitivity to such decays puts the proton lifetime as greater than about 10^{34} years [Abe *et al.* (2017); Takenaka *et al.* (2020)] from the Super-Kamiokande experiment, corresponding to an ultraviolet scale $M \sim 10^{15}$ GeV. Next generation experiments, e.g. using the Hyper-Kamiokande detector, will push the sensitivity towards a possible lifetime about 10^{35} years [Bian *et al.* (2022)].

5.3.3   *The Pauli term in QED and the eEDM*

If we treat QED as an effective theory, then we might think to supplement the Lagrangian, Eq. (2.1), with the additional gauge invariant but non-renormalisable Pauli term

$$\mathcal{L}^{\text{Pauli}} = i(e/2M)\,\bar{\psi}(\gamma^{\mu}\gamma^{\nu} - \gamma^{\nu}\gamma^{\mu})\psi\, F_{\mu\nu}. \tag{5.9}$$

The Pauli term is a $D = 5$ extension of QED that comes suppressed by some large mass scale M which represents the ultraviolet limit of the effective theory. If present, this term induces a contribution to the electron's anomalous magnetic moment a_e of size $4e/M$ in addition to the perturbative contributions in Eqs. (2.10)–(2.12) [Weinberg (1995)]. If one takes the Pauli term as parametrising any extra new physics contributions, then precision measurements of the eEDM [Fan *et al.* (2023)] give $M > 4 \times 10^9$ GeV. Inclusion of the $D = 5$ Pauli term can remove the Landau pole singularity in the QED running coupling [Djukanovic *et al.* (2018)].

5.3.4   *Axions and the strong CP puzzle*

Going beyond the fields of the Standard Model, one key idea to resolve the strong CP puzzle — the absence of CP violation that arises with any non-zero QCD Θ_{QCD} angle and induced by non-perturbative gluon topology together with finite quark masses — involves possible new very light mass pseudoscalar particles called axions [Peccei and Quinn (1977); Weinberg (1978); Wilczek (1978)]. The axion would be the Nambu–Goldstone boson associated with a new spontaneously broken anomalous $U(1)_{\text{PQ}}$ Peccei–Quinn symmetry [Peccei and Quinn (1977)].

If present, the axion particle a would couple through the Lagrangian term

$$\mathcal{L}_a = -\frac{1}{2}\partial_{\mu}a\partial^{\mu}a + \left[\frac{a}{M} - \Theta_{\text{QCD}}\right]\frac{\alpha_s}{8\pi}G_{\mu\nu}\tilde{G}^{\mu\nu} - \frac{if_{\psi}}{M}\partial_{\mu}a\,\bar{\psi}\gamma_5\gamma^{\mu}\psi - \cdots, \tag{5.10}$$

where the term in ψ denotes possible fermion couplings to the axion a and $f_{\psi} \sim \mathcal{O}(1)$. Here, the mass scale M plays the role of the axion decay constant and sets the scale for this new physics. The axion transforms under the new global U(1) Peccei–Quinn symmetry to cancel the Θ_{QCD} term, with strong CP violation replaced by the axion coupling to gluons (and also photons). The axion here develops a vacuum expectation value with the potential minimised at $\langle\text{vac}|a|\text{vac}\rangle/M = \Theta_{\text{QCD}}$. The mass squared of the QCD axion is given by

$$m_a^2 = \frac{F_\pi^2}{M^2} \frac{m_u m_d}{(m_u + m_d)^2} m_\pi^2. \tag{5.11}$$

Here, m_π and F_π are the pion mass and decay constant, and m_u and m_d the up and down quark masses. Note that the axion coupling to Standard Model particles and its mass enter at $D = 5$ with the $1/M$ suppression factor. Thus, the "axion solution" to the strong CP puzzle involves a delicate interplay of $D = 4$ and $D = 5$ physics.

Axion searches are a vigorous topic of experimental research [Döbrich (2022); Rosenberg *et al.* (2024)]. Constraints from experiments tell us that M must be very large. Laboratory based experiments based on the two-photon anomalous couplings of the axion [Ringwald (2015)], ultracold neutron experiments to probe axion to gluon couplings [Abel *et al.* (2017)], together with astrophysics and cosmology constraints suggest a favoured mass for the possible QCD axion between $1\mu eV$ and 3 meV [Baudis (2018); Kawasaki and Nakayama (2013)], corresponding to M between about 6×10^9 and 6×10^{12} GeV. The small axion interaction strength, $\sim 1/M$, means that the small axion mass corresponds to a long lifetime and stable dark matter candidate, e.g. lifetime longer than about the present age of the Universe. If the axions were too heavy they would carry too much energy out of supernova explosions, thereby observably shortening the neutrino arrival pulse length recorded on Earth in contradiction to Sn 1987a data [Kawasaki and Nakayama (2013)].

Beyond possible QCD axions, one can also consider possible "axion-like" particles, ALPs. These might come with similar $1/M$ type couplings to Standard Model particles (especially the photon) plus an extra ALP mass term beyond the QCD axion mass contribution in Eq. (5.11) [Hook (2019)]. In general, possible ALP particles do not necessarily resolve the QCD strong CP puzzle with the answer depending on their interactions. Depending on the origin and size of the extra mass term, such particles can involve both $D = 5$ physics through coupling to Standard Model particles and $D = 4$ physics through the new mass term (which needs extra explanation as a contribution beyond QCD and beyond the BEH mass terms in the Standard Model Lagrangian). As such, their dynamics may lie beyond a pure $D = 5$ contribution in the low energy expansion.

In the Standard Model, the flavour structure of the Yukawa couplings is not controlled by the dynamics unless one imposes the constraint from Higgs vacuum stability on the value of m_t. There are ideas that these couplings might be controlled by a (global or local) symmetry which gets spontaneously broken at some large scale by the vacuum expectation value of a "flavon" field — the Froggatt–Nielsen mechanism [Froggatt and Nielsen (1979)]. The axion is a good candidate for this "flavon" field in which case the axion might be closely connected with flavour physics — for recent discussion see [Pokorski (2023)].

5.3.5   *Summary*

With the Standard Model gauge symmetries taken as emergent and dynamically generated, the full theory is not the physics truncated to mass dimension four operators but also includes an infinite tower of higher dimensional terms suppressed by powers of the large emergence scale, see Table 5.1. At low energies the physics is determined by a relatively small number of operators with mass dimension at most four. For these specific terms gauge invariance and renormalisability restrict the number of possible operator contributions and strongly constrain the global symmetries of the system. Extra global symmetry breaking terms can occur in higher dimensional operators which enter the action suppressed by powers of the large scale of emergence. Examples are the L_5 lepton number violating operator, the QED Pauli term and axion-like couplings at $D = 5$, plus the baryon number violating operator B_6 and new CP violating terms at $D = 6$.

Table 5.1.   Typical operators in a low energy expansion. The large ultraviolet scale M here is taken to be about 10^{16} GeV. The operators at $D = 1, 2, 3$ are tamed by the symmetries. Table adapted from [Jegerlehner (2014a)].

	dimension	operator	scaling behaviour
	.	∞–many	
	.	irrelevant	
$\uparrow$	.	operators	
no			
data	$D = 6$	$B_6, (\Box\phi)^2, (\bar{\psi}\psi)^2, \ldots$	$(E/M)^2$
		$D = 5$	$L_5, \bar{\psi}\sigma^{\mu\nu}F_{\mu\nu}\psi, a\frac{\alpha_s}{8\pi}G_{\mu\nu}\tilde{G}^{\mu\nu}, \ldots$	(E/M)
experimental			
data	$D = 4$	$(\partial\phi)^2, \phi^4, (F_{\mu\nu})^2, \bar{\psi}i\gamma^\mu D_\mu\psi, \ldots$	$\ln(E/M)$
$\downarrow$			
	$D = 3$	$\phi^3, \bar{\psi}\psi$	(M/E)
	$D = 2$	$\phi^2, (A_\mu)^2$	$(M/E)^2$
	$D = 1$	ϕ	$(M/E)^3$

These higher dimensional terms become active in the particle dynamics when we are sensitive to energy scales close to the large emergence scale. This is here taken around 10^{16} GeV which is within the range where both λ and β_λ might touch zero. At laboratory energies the higher dimensional terms will be manifest through tiny masses and couplings. The extension of this discussion to cosmology is given in Chapters 9 and 11 with focus on the cosmological constant and dark matter respectively.

Chapter 6

Emergent Gauge Symmetries in Quantum Condensed Matter Systems

Emergent gauge symmetries are important in quantum condensed matter systems with long range entanglement and topological order, so called topological phases of matter. Besides fermion and spin-zero boson quasiparticles, one can also have extra gauged spin-one boson excitations. The accompanying emergent gauge symmetries arise in topological phase transitions which occur without a local order parameter. The emergent gauge fields are coupled to but distinct from the underlying electromagnetic interactions between electrons and atoms. For general reviews, see [Moessner and Moore (2021); Powell (2020); Zaanen and Beekman (2012)]. With emergent gauge symmetries we make symmetry instead of breaking it.

In condensed matter physics emergent gauge symmetries were first found in the low energy limit of the Fermi–Hubbard model of electron systems by Anderson and collaborators [Affleck et $al.$ (1988); Baskaran and Anderson (1988)]. This model is important in discussions of high temperature superconductors and quantum spin liquids [Powell (2020); Sachdev (2016, 2019); Spałek et $al.$ (2022)], as well as in quantum simulations of gauge theories [Bañuls et $al.$ (2020); Banerjee et $al.$ (2012); Zohar et $al.$ (2016)]. The Fermi–Hubbard model with a two dimensional lattice serves as an important and useful prototype model to explain the key concepts and ideas. In this chapter, we first review how emergent gauge symmetries arise in this system. With particle physics in mind, we then describe work by Volovik on the A-phase of superfluid ^{3}He where the quasiparticles include gapless chiral fermions coupled to emergent SU(2) plus U(1) gauge fields [Volovik (2003, 2008)]. With ^{3}He-A one also finds emergent Lorentz invariance with limiting velocities and emergent gravity. Finally, we briefly discuss $3+1$ dimensional string-nets which involve qubit chains in a lattice environment [Levin and Wen (2005a, 2005b); Wen (2002, 2004)]. These chains can condense with excitations providing a model for electrons and photons as emergent degrees of freedom which can be extended to include quarks and gluons. Here one finds emergent gauge fields plus fermions linked to details of long range quantum entanglement. Other examples of emergent gauge symmetries include spin ice in magnetic systems with emergent Maxwell theory and magnetic monopoles [Rehn and Moessner (2016)], the quantum Hall effect [von Klitzing *et al.* (1980)] and its description in terms of a topological field theory with an emergent gauge field (for a review, see e.g. [Tong (2016)]), as well as various simple quantum mechanical systems [Wilczek and Zee (1984)].

Besides ³He-A, emergent Lorentz invariance is also observed in the infrared limit of other many-body quantum systems starting from a non-relativistic Hamiltonian. Here, though, some fine tuning [Levin and Wen (2006b)] or extra symmetry [Volovik (2003)] may be needed to ensure the same effective limiting velocity c for all species of (quasi-)particles in the infrared emergent theory.

Long range entanglement [Kitaev and Preskill (2006); Levin and Wen (2006a)] is a universal property of quantum many-body systems and insensitive to microscopic details of the Hamiltonian. The following dancing analogy [Wen (2013)] is helpful to understand the difference between topological order and symmetry-breaking with an associated local order parameter. First, with symmetry-breaking orders, every particle/spin, or every pair of particles/spins, dances by itself and they all dance in the same way (corresponding to the long-range order). With topological order one has a global dance, where every particle/spin is dancing with every other particle/spin in a collective and very organised fashion, not just fermions dancing in pairs. This extra organisation involves the decoupling of potential degrees of freedom, like photons have transverse but not longitudinal polarisations. Global dancing patterns are a collective effect produced from various local dancing rules which corresponds to a pattern of long range entanglement. Transformations between states with different long range entanglement patterns are described by topological phase transitions. In the context of quantum computers, with long range entanglement the time taken for disentanglement depends on the size of the system whereas for short range entanglement it is size independent. More formal definition in terms of unitary evolution and Hilbert space constructions is given in [Chen *et al.* (2010); Zeng *et al.* (2019)].

6.1 The Fermi–Hubbard Model and Its Low Energy Limit

The Fermi–Hubbard model describes electron behaviour in an atomic lattice. One considers a lattice of atoms supporting only a single atomic state, which can hold up to two electrons with opposite spins. The electrons interact with the potential of a static lattice of ions. One neglects any motion of the ion lattice, being only interested in interactions of electrons and not in dynamical lattice effects, such as phonons. The electrons interact via Coulomb repulsion. Further, one assumes that all except the lowest band have very high energies and are, thus, energetically unavailable. Also, that the remaining band has rotational symmetry. The electron hopping matrix, which describes electron motion from one lattice site to another, depends just on the distance between lattice sites. Finally, one restricts the model to nearest neighbour interactions (with underlying matrix elements decreasing fast with increasing distance).

The Fermi–Hubbard model Hamiltonian then has two terms: a hopping term between nearest neighbour sites with coupling strength t, plus an "on-site" Fermi–Hubbard repulsion term U,

$$\mathcal{H} = -t \sum_{(ij)\sigma} c_{i\sigma}^{\dagger} c_{j\sigma} + U \sum_{i} c_{i\uparrow}^{\dagger} c_{i\uparrow} c_{i\downarrow}^{\dagger} c_{i\downarrow}. \tag{6.1}$$

Here, a square lattice is assumed where ij are nearest neighbour bonds. $c_{i\sigma}^{\dagger}$ and $c_{i\sigma}$ are the creation and annihilation Fock operators for electrons with spin $\frac{1}{2}$ on site i. The first term prefers non-localisation whereas the second prefers localisation with just one electron on each lattice site. Extra "doping" terms can be included by adding a chemical potential.

In the low-energy Mott limit $U \gg t$ the Fermi–Hubbard system behaves as an insulator. (With extra doping terms described by adding a chemical potential, it becomes a model for describing high temperature superconductors.) Treating the hopping term as a perturbation and keeping the leading term evaluated using Rayleigh–Schrödinger perturbation theory, the Fermi–Hubbard model reduces to the Heisenberg magnet Hamiltonian. For the half filled system one finds

$$\mathcal{H}_{\text{eff}} = J \sum_{i,j} (c_{i\alpha}^{\dagger} \sigma_{\alpha\beta} c_{i\beta}) \cdot (c_{j\alpha}^{\dagger} \sigma_{\alpha\beta} c_{j\beta}), \tag{6.2}$$

where $J = 4t^2/U$, the σ denote SU(2) Pauli matrices, and one has the constraint $c_{j\alpha}^{\dagger} c_{j\alpha} = 1$.

The strongly correlated electron system can be described using a slave-particle representation using auxiliary fermion and boson operators. One writes the c-electron Fock operators as a combination of "spinon" f-electrons carrying spin and no electric charge, and spinless "holons" which carry the electric charge, that is, with spin-charge separation; for a review see [Frésard (2015)]. In the low-energy Fermi–Hubbard model with half filling the spin operators appearing in the product in Eq. (6.2) are chargeless and the low energy system can be written just in terms of the f-electrons, $c \mapsto f$ in Eq. (6.2).

The Hamiltonian (6.2) has the important local gauge symmetry $f_{j\sigma}^\dagger \to e^{i\theta_j} f_{j\sigma}^\dagger$. The Fermi–Hubbard system exhibits entanglement and quantum correlations in its ground state in the $U \gg t$ limit [Powell (2020)].[1]

The Hamiltonian Eq. (6.2) can also be expressed in the form [Baskaran and Anderson (1988)]

$$\mathcal{H}_{\text{eff}} = J \sum_{ij} b_{ij}^\dagger b_{ij}, \qquad (6.3)$$

where $b_{ij}^\dagger = (1/\sqrt{2})(f_{i\uparrow}^\dagger f_{j\downarrow}^\dagger - f_{i\downarrow}^\dagger f_{j\uparrow}^\dagger)$ are bosonic single (two electron) creation and annihilation operators, which are important in the RVB theory of superconductivity [Anderson *et al.* (1987); Baskaran *et al.* (1987)]. These two electron excitations might form a Bose–Einstein condensate. Through Elitzur's theorem the thermal average $\langle b_{ij} \rangle = 0$ at all temperatures.

The model system Eq. (6.2) also exhibits a local SU(2) gauge symmetry. To see this, first consider the electron operators (f_1, f_2) and $(f_2^\dagger, -f_1^\dagger)$ which transform as SU(2) spin doublets. These are combined to form the matrix

$$\Psi = \begin{pmatrix} f_1 & f_2 \\ f_2^\dagger & -f_1^\dagger \end{pmatrix} \qquad (6.4)$$

which transforms under global SU(2), $\Psi \to \Psi g$. One can define a second local SU(2) symmetry by $\Psi \to h\Psi$. Here, g and h denote SU(2) rotations, viz. $e^{i\vec{\sigma}.\vec{\omega}/2}$ where $\vec{\sigma}$ denotes the SU(2) Pauli matrices and $\vec{\omega}$ is spacetime independent for g and spacetime dependent for h. Spin operators for

---

[1]The emergent U(1) gauge field here is confined for all values of the coupling in two and three dimensional systems with the Wegner–Wilson loop satisfying an "area law" for large loops [Baskaran and Anderson (1988)].

global SU(2) can be written $S = \frac{1}{2}\Psi^\dagger\Psi\sigma^T$ where σ^T is the transpose of σ. Since $\Psi^\dagger \to g^\dagger\Psi^\dagger h^\dagger$ it follows that the spin operators are invariant under local SU(2). That is, the Heisenberg interaction is invariant under local SU(2) gauge transformations with h denoting an element of the gauge group.

The Hamiltonian in Eq. (6.2) can be written in terms of the spin operators as

$$\mathcal{H}_{\text{eff}} = J/4 \sum_{i,j}(\text{tr }\Psi_i^\dagger\Psi_i\sigma^T)\cdot(\text{tr }\Psi_j^\dagger\Psi_j\sigma^T). \tag{6.5}$$

The local gauge symmetry within the Heisenberg model acts trivially on the spin operators but becomes interesting within the large U limit of the Fermi–Hubbard model with electron operators at half filling. This is a consequence of the redundancy of parametrising spin operators by electron operators. Note that it is the "spinon" f-electrons that feel the emergent SU(2) gauge symmetry here rather than it being a property of the charged c-electrons of QED which appear in the more general Fermi–Hubbard Hamiltonian Eq. (6.1).

The emergent gauge symmetry seen here comes with an energy barrier. Local SU(2) gauge invariance is valid up to below the Mott–Hubbard energy gap. For large but finite U there is an approximate gauge symmetry in the sense that it is only broken in the sector of the Hilbert space containing high energy states with energies of order U.

Adding in a new spin one gauge field gives the net Lagrangian

$$\mathcal{L} = \frac{1}{2}\sum_j \text{tr }\Psi_j^\dagger\left(i\frac{\partial}{\partial t} + B_j\right)\Psi_j - \mathcal{H}_{\text{eff}} \tag{6.6}$$

with the gauge field $B = \frac{1}{2}\sigma\cdot\mathbf{B}$ transforming as $B \to h(B + i(\partial/\partial t))h^\dagger$ under the local SU(2) gauge transformations associated with h. The three components of $\mathbf{B}$ act as Lagrange multipliers and guarantee the half filled system with constraint of one particle per site [Affleck *et al.* (1988)].

We refer to [Sachdev (2019)] for detailed discussion of the phase diagram of the Fermi–Hubbard model including confinement and Higgs phases, as well as application to high temperature superconductors. Beyond the strongly correlated electron interactions, phononic vibrations of the atomic lattice exhibit bosonic statistics independent of the fermionic or bosonic nature of the atoms on the lattice sites [Moessner and Moore (2021)].

6.2 Fermi Points and Emergent Gauge Symmetries in Superfluid ^{3}He-A

The low temperature A-phase of superfluid ^{3}He exhibits similar structure to the particle physics Standard Model. One finds emergent gauge and gravity fields. The quasiparticles involve gapless chiral fermions associated with Fermi points in momentum space interacting through emergent SU(2) and U(1) gauge bosons [Volovik (2003, 2008, 2013)]. The quasiparticles also exhibit limiting velocities and the theory behaves like a relativistic quantum field theory.

Quasiparticle properties in condensed matter systems correspond to universality classes characterised by topology in three-dimensional momentum space. The system may be fully gapped, it may develop the Fermi surface or, alternatively, one can find singular points in momentum space called Fermi points. Without using the microscopic theory, and only from topology in the momentum space, one can predict all possible types of behaviour of the condensed matter many-body system at low energy that do not depend on details of atomic structure.

Systems with a gap in their spectrum,

$$E^2 = v_F^2 (p - p_F)^2 + \Delta^2, \tag{6.7}$$

include semiconductors and normal superconductors as well as quasiparticles in the B-phase of ^{3}He.

Gapless systems with a Fermi surface include Landau–Fermi liquids such as normal liquid ^{3}He and normal metals. Here, the quasiparticle energies E are related to the 3-momentum p by

$$E = v_F (p - p_F), \tag{6.8}$$

where p_F is the Fermi momentum and $v_F = p_F/m$ is the corresponding Fermi velocity. The Fermi surface is stable because it is a topological defect — a quantised vortex in momentum space.

Superfluid ^{3}He-A falls into the third category. Here, the quasiparticles involve gapless chiral fermions associated with Fermi points in momentum space interacting through emergent SU(2) and U(1) gauge bosons [Volovik (2003, 2008)]. The quasiparticle theory behaves like a relativistic quantum field theory. That is, the low temperature A-phase of superfluid ^{3}He exhibits some similar structure to the particle physics Standard Model. Fermi points are topologically stable point nodes in momentum space.

In addition, there are also systems with topologically unstable lines of nodes called Fermi lines which become stable only if some special symmetry is obeyed, e.g. in high temperature superconductors with d-state Cooper pairing [Volovik (2007)].

In low temperature physics ^{4}He becomes a superfluid at 2K whereas ^{3}He remains as a normal liquid at these temperatures and becomes superfluid only at 2.6 mK with a much richer phase diagram [Lee (1997); Osheroff (1997); Vollhardt and Wolfle (2000); Volovik (2003)]. The ^{3}He becomes a superfluid around 2 mK and pressure less than 34 bars (above which it becomes a solid). Whereas the Cooper pairs in ^{4}He are in s-wave with $S = l = 0$, the Cooper pairs for fermionic ^{3}He come in p-wave (meaning nine order parameters) introducing some anisotropy in the physics of the superfluid.

Without an external magnetic field there are two phases of superfluid ^{3}He. The B-phase involves Cooper pairs with $\mathbf{J} = \mathbf{l} + \mathbf{S} = \mathbf{0}$. Angular momentum $J = 0$ means that the quantum vacuum of the B-phase is isotropic under simultaneous rotations in spin and coordinate space, and thus ensures the isotropy of the liquid. In the B-phase, the energy gap is isotropic just like for s-wave Cooper pairs but the system is intrinsically anisotropic. The B-phase forms at 1.8 mK. The Cooper pairs here are a linear combination of the three spin substates $|\uparrow\uparrow\rangle$, $\frac{1}{\sqrt{2}}\{|\uparrow\downarrow\rangle + |\downarrow\uparrow\rangle\}$ and $|\downarrow\downarrow\rangle$ [Lee (1997)].

The A-phase forms at elevated pressure, about 21 bars, where spin fluctuations cause the pair correlations in the condensed states to change the pairing interaction between ^{3}He quasiparticles. "Strong coupling effects" enter beyond weak coupling theory (where the B-phase always has the lowest energy). One finds a new phase ^{3}He-A with the property that, in contrast with ^{3}He-B, its magnetic suspectibility is essentially the same as that of the normal liquid. Thus, in this phase the spin substate with $S_z = 0$ — the only one that can be reduced appreciably by an external magnetic field — is absent. ^{3}He-A involves just $|\uparrow\uparrow\rangle$ and $|\downarrow\downarrow\rangle$ Cooper pairs with no $S_z = 0$ state and forms at 2.6 mK and 21 bars of pressure.

A third A_1-phase lies between the A and B phases and involves just $|\uparrow\uparrow\rangle$ Copper pairs. The A_1 phase is only stable in the presence of an external magnetic field.

In the A-phase, the energy gap for the fermionic quasiparticles comes with angular dependence [Anderson and Brinkman (1973); Anderson and Morel (1961)]. For fermion quasiparticles with energy E, 3-momentum p and mass m one finds

$$E^2(\mathbf{p}) = \left(\frac{\mathbf{p}^2}{2m} - \mu\right)^2 + c_\perp^2 (\mathbf{p} \times \hat{\mathbf{l}})^2 \qquad (6.9)$$

corresponding to the Nambu–Bogoliubov Hamiltonian [Volovik (2007)]. Here, the unit vector $\hat{l}$ corresponds to the direction of the orbital momentum of the Cooper pair; $c_\perp$ is the speed of the quasiparticles if they propagate in the plane perpendicular to $\hat{l}$ and μ is the chemical potential. When $p = p_F$ with Fermi momentum $p_F = \sqrt{2m\mu}$ the gap is $\Delta(\vartheta) = p_F c_\perp \sin \vartheta$ for the p-wave state in Eq. (6.9), where ϑ is the polar angle. For Cooper pairs forming with positive chemical potential $\mu > 0$ there are two points at the former Fermi surface in three-dimensional momentum space with $E(\mathrm{p}) = 0$ corresponding to nodes in the spectrum of quasiparticles. This is instead of a Fermi surface without the angular dependent term.[2] These so called Fermi points occur at $\mathrm{p}_1 = p_F \hat{l}$ and at $\mathrm{p}_2 = -p_F \hat{l}$ with Fermi momentum $p_F = \sqrt{2m\mu}$. The Fermi velocity $v_F = p_F/m = c_\parallel$ represents the maximum speed of quasiparticles propagating parallel to $\hat{l}$. For superfluid ^{3}He-A $c_\perp/c_\parallel \approx 10^{-5}$ [Volovik (2003)].

The direction of the relative orbital angular momentum $\hat{l}$ is the same for all Cooper pairs. Likewise, the anisotropy axis of the spin part of the Cooper pair wavefunction $\hat{d}$ has the same fixed direction in every pair. (More precisely $\hat{d}$ is the direction along which the total spin of the Cooper pair vanishes $\hat{d} \cdot \mathbf{S} = 0$.) Therefore, in the A phase the anisotropy axes $\hat{d}$ and $\hat{l}$ of the Cooper-pair wave function are long-range ordered. That is, they are preferred directions in the whole macroscopic sample. This result implies a pronounced anisotropy of the A-phase in all of its properties.

Quasiparticle excitations in the neigbourhood of Fermi points are Weyl fermions which interact like a relativistic quantum field theory with emergent gauge bosons [Volovik (2003, 2013)]. The key physics involves the momentum space topology of the fermion quasiparticles. To understand this physics and emergent gauge symmetries here, these are associated with the Fermi point and the freedom in choosing the position of this Fermi point on the former Fermi surface. To see this consider the fermion quasiparticle propagator $\mathcal{G}$ in the vicinity of the Fermi point $p = p^{(0)}$, viz.

$$\mathcal{G}^{-1}(p_0) = i\omega - H(\mathrm{p}) = e_i{}^k \Gamma^i (p_k - p_k^{(0)}) + \text{higher order terms.} \quad (6.10)$$

Here, $\Gamma_i = (1, \sigma_i)$ and the matrix e_i^k is the analogue of the dreibein with $g^{ik} = e_j^i e_j^k = \mathrm{diag}(c_\perp^2, c_\perp^2, c_\parallel^2 = p_F^2/m^2)$ playing the role of an effective dynamical metric in which fermions move along geodesic lines.

---

[2]In ideal Fermi gases the Fermi surface at $p = p_F = \sqrt{2\mu m}$ is the boundary in p-space between the occupied states ($n_p = 1$) at $p^2/2m < \mu$ and empty states ($n_p = 0$) at $p^2/2m > \mu$.

The Green's function has a singularity at the Fermi point. "Higher order terms" represent additional contributions slightly away from the Fermi point. At the Fermi points the quasiparticle propagators are characterised by the topological invariant

$$N_3 = \text{tr}\mathcal{N}, \quad \mathcal{N} = \frac{1}{24\pi^2}\epsilon_{\mu\nu\lambda\gamma} \int_S d\sigma^\gamma \, \mathcal{G}\partial_{p_\mu}\mathcal{G}^{-1}\mathcal{G}\partial_{p_\nu}\mathcal{G}^{-1}\mathcal{G}\partial_{p_\lambda}\mathcal{G}^{-1}. \quad (6.11)$$

Taking the Fermi point position as $p_k^{(0)} = 0$, the propagator perceived by an internal local observer simplifies to

$$\mathcal{G}^{-1}(p_0) = ip_0 + N_3 \, c\sigma \cdot \mathrm{p} + \text{ higher order terms.} \quad (6.12)$$

At the propagator singularity the Hamiltonian then becomes $H = N_3 \, c\sigma.\mathrm{p}$. Chirality emerges as a property of the low energy system. It is determined by the topological invariant N_3. Fermions are right-handed if the determinant of the matrix e_j^i is positive, which occurs at $N_3 = +1$; the fermions are left-handed if the determinant of the matrix e_j^i is negative which occurs at $N_3 = -1$. The Fermi point behaves as a hedgehog in momentum space with plus/minus signs $N_3 = \pm 1$ corresponding to the spins pointing out/in.

Invariance under changes in the position of the Fermi point correspond to effective gauge symmetries. The topological numbers $N_3 = \pm 1$ correspond to an emergent U(1) with Hamiltonian and energy

$$H \to e_k^i \sigma^k (p_i - eA_i), \quad E^2(\mathrm{p}) \to g^{ik}(p_i - eA_i)(p_k - eA_k) = 0. \quad (6.13)$$

Here, $\mathrm{A} = p_F \hat{\mathrm{l}}$ the effective gauge field. The "electric charge" is either $e = +1$ or -1 depending on the Fermi point. The topological numbers $N_3 = \pm 2$ correspond to emergent local SU(2). Under the most general transformation one finds

$$H = +c\sigma \cdot p \to e_i^{\ k}\Gamma^i \cdot (p_k - p_k^{(0)})$$
$$E^2 \to g^{\mu\nu}(p_\mu - eA_\mu - g\tau.W_\mu)(p_\nu - eA_\nu - g\tau.W_\nu) = 0.$$
$$(6.14)$$

This expression for the fermion quasiparticle's energy squared, E^2, corresponds to emergent relativistic behaviour. The gauge symmetries correspond to properties of the leading linear term in the Hamiltonian very close to the Fermi points. The quasiparticles of ^{3}He-A (fermions and gauge fields) also exhibit an analogue of the axial anomaly with the experimental observation discussed in [Bevan *et al.* (1997)]. There is a further interesting application to the cosmological constant. For a droplet of the ^{3}He-A quantum liquid, the net vacuum energy or effective "cosmological constant" vanishes up to boundary surface corrections [Volovik (2003)]. One finds an effective gravity corresponding to variations in the metric g^{ij} which is emergent in the ^{3}He-A system together with the emergent SU(2) and U(1) gauge symmetries and the chiral fermionic Weyl quasiparticles. The corresponding "gravitons" here would come with quasiparticle structure and dissolve along with the emergent fermionic and $J = 1$ gauge boson quasiparticles when one goes through the emergence transition corresponding to the formation of ^{3}He-A.

The superfluid ^{3}He-A system with emergent massless chiral fermions coupled to SU(2) and U(1) gauge fields looks similar to the Standard Model in a possible symmetric phase that might exist above any electroweak phase transition. Within the ^{3}He-A system if the Fermi points associated with left-handed and right-handed states coincide, then the topological charges annihilate at so called "marginal Fermi points". These states then mix yielding massive Dirac type fermions [Volovik (2007)].

Summarising, in ^{3}He-A fermion quasiparticles are gapped except close to two Fermi points

$$E = \pm N_3 \, c\sigma.p \qquad (6.15)$$

corresponding to emergent chiral fermions with limiting velocity c and chirality ± 1. The ordinary spin degree of freedom is perceived by an inner observer as a local SU(2) gauge symmetry, like weak isospin in particle physics. This gauged SU(2) symmetry is dynamical as it represents some collective motion of the fermionic vacuum. The emergent gauge symmetry appears only in the low energy limit. The fermion quasiparticles are related by global discrete symmetries. Since the action for the gauge bosons is obtained by integrating over the fermions, one obtains the same limiting velocity and emergent Lorentz invariance for each species of quasiparticle [Volovik (2003)].

6.3 String-net Condensation

In $3 + 1$ dimensions string-net condensation has been proposed as a topological phases model for emergent electrons and photons, as well as for quarks and gluons. String-net condensed states are liquids of fluctuating networks of strings. A condensate is formed from these extended structures rather than particles. String excitations above this condensate correspond to massless spin-one particles and the ends of the strings to fermions [Levin and Wen (2005a, 2005b); Wen (2002, 2004)].

The idea starts with a lattice structure with spin zero bosons – qubits. These can form a string structure with occupied sites building on an empty "vacuum". Condensation of these strings can occur as separate to condensation of individual bosons. One introduces an energy penalty for strings that end or change type in empty space. Strings then arrange themselves to minimise the energy into effective extended objects.

The general Hamiltonian for the string-net system consists of a sum over kinetic, potential and constraint terms, with the latter imposing large energy penalties for "illegal" string branchings. When the potential term dominates the kinetic term then the ground state involves few string-nets. When the kinetic term dominates the potential term, then one expects a ground state with many fluctuating string-nets with typical length of order the size of the system. These may then "condense" in a quantum phase transition between the two ground states when the ratio of the kinetic to potential terms is of order one.

This string condensation is beyond any condensation of individual boson sites into a Bose–Einstein condensate. Excitations above the condensate are gauge bosons and fermions. Closed strings correspond to spin one gauge bosons and the ends of open strings correspond to spin $\frac{1}{2}$ fermions satisfying Fermi–Dirac statistics.[3] Strings can vibrate in two space dimensions. One obtains two physical degrees of freedom, just as there are two transverse photon polarisations. In this picture light waves become collective excitations of the string-nets and electrons correspond to one end of string. The emergent QED satisfies its own version of Gauss's Law equivalent to electric flux conservation through the strings. The string-net picture thus gives a unification of gauge symmetry and spin-statistics. With string-net condensation, gauge theory becomes a theory for long range entanglement [Zeng *et al.* (2019)]. One prediction is that all fermionic excitations must carry some gauge charges. Fermions are the topological defects of long range entanglement.

This picture can be generalised to emergent Yang–Mills theory with non-abelian SU(3) gauge degrees of freedom and different condensation

---

[3]One finds bosons as the end of strings summed over all configurations. Fermions arise when summed over strings weighted by the number of crossings when the three-dimensional strings are projected onto a two-dimensional plane.

"patterns" corresponding to different gauge groups [Levin and Wen (2005a)]. Chiral fermion degrees of freedom remain a more open issue here, perhaps related to issues of how to include chiral fermions on the lattice. Further interesting questions are how to extend this theory to include renormalisation group flow with running couplings and related connections to confinement and Higgs phenomena plus topological vacuum structure for emergent Yang–Mills theories.

In condensed matter physics in $2 + 1$ dimensions string-nets may be important for understanding quantum spin liquids. The mineral Herbertsmithite with Kagome lattice structure was discovered to exhibit properties of a quantum spin liquid [Han *et al.* (2012)] (whether gapped or gapless) with possible string-net like structure; for recent discussion see [Zeng *et al.* (2019)].

Chapter 7

Gravitation and General Relativity

So far, we have focused on the particle physics Standard Model. Gravitation is described in our present experiments by Einstein's theory of General Relativity. Whereas the Standard Model is a quantum field theory, General Relativity is purely classical. It can also be described as a gauge theory (under local coordinate transformations instead of local rotations of the phases of the fields).

In this chapter, we describe General Relativity and recent observations of its properties including the equivalence principle, black holes and the discovery of gravitational waves. In Chapter 8, this discussion is extended to cosmology where observations tell us that normal Standard Model baryonic matter comprises just 5% of the energy budget of the Universe. For the rest, 68% of the energy budget is in the dark energy that drives the accelerating expansion of the Universe and which is believed to be the same everywhere with uniform density. The remaining 27% is in dark matter that clumps like normal Standard Model matter under gravitational attraction but which is non-luminous. So far, the evidence for dark matter comes only through its gravitational interaction. The challenge to understand dark energy with connections to vacuum energy in quantum field theories and dark matter are prime topics of investigation at the particle-gravity frontier. A further interesting question is whether gravitation might be emergent — perhaps along with the Standard Model — similar to the situation with quasiparticle interactions in ^{3}He-A.

7.1   General Relativity

General Relativity is Einstein's dynamical theory of gravity where gravitation is connected with the geometry of four-dimensional spacetime

[Dirac (1996); Einstein (1956); Misner *et al.* (1973); Straumann (2013); Weinberg (1972)]. General Relativity is based on and at the same time extends Newton's theory of gravitation:

(1) Space and time form a four-dimensional manifold, called spacetime. Its geometrical properties are described by its metric $g_{\mu\nu}$. The metric is determined by gravitation, and the components are calculated from Einstein's equations, which are explained below. The metric may depend on the location in space and time.
(2) The physical laws are the same in every coordinate scheme; mathematically speaking, they have to be covariant. (This condition leads to a gauge theory interpretation of gravitation.) Invariance of the laws of physics under arbitrary differentiable coordinate transformations — also called diffeomorphism invariance — means that spacetime coordinates are properties of our description of Nature rather than properties of Nature itself.
(3) Locally the gravitational field can be (almost) transformed away by choosing a coordinate system in which the metric $g_{\mu\nu}$ is flat. In such "local inertial systems" the laws of Special Relativity are valid.

These principles lead to Einstein's equations where gravitation and spacetime curvature couple to the energy-momentum tensor. Quoting John Wheeler (page 235 of [Wheeler and Ford (1998)]): *Spacetime tells matter how to move; matter tells spacetime how to curve.* Just like with the gauge theories of the particle physics Standard Model, the local symmetry properties of the theory determine the dynamics.

Building theories of gravity and General Relativity a vital ingredient is the equivalence principle [Charlton *et al.* (2020); Weinberg (2008); Will (2014)]. The equivalence principle enters at three levels. Weak or Galilean equivalence, the WEP, concerns the universality of free fall, that the inertial and gravitational masses should be equal. The Einstein equivalence principle, EEP, involves the extension to General Relativity type theories and says that the outcome of non gravitational experiments in free fall should be independent of the velocity of the apparatus as well as its place in spacetime. The EEP includes the following postulates (1) WEP, (2) Local Lorentz invariance (LLI): the outcome of any local nongravitational experiment conducted in free fall is independent of the velocity and the orientation of the apparatus and (3) Local position invariance (LPI).

Strong equivalence, the SEP, concerns the extension from test masses to self gravitating bodies. With the SEP gravitation proceeds by minimal coupling via the spacetime connection defined below, there are no fifth forces from extra scalar gravitational interactions and the value of Newton's constant G is independent of where it is measured in spacetime.

General Relativity is a gauge theory built on invariance under local coordinate transformations

$$x^\mu \to \tilde{x}^\mu(x). \tag{7.1}$$

For a general tensor

$$S^\mu{}_{\nu\rho} \to S^{\mu'}{}_{\nu'\rho'} = \frac{\partial \tilde{x}^{\mu'}}{\partial x^\mu} \frac{\partial x^\nu}{\partial \tilde{x}^{\nu'}} \frac{\partial x^\rho}{\partial \tilde{x}^{\rho'}} S^\mu{}_{\nu\rho}. \tag{7.2}$$

If an equation between two tensors holds in one frame, it holds in all frames. To formulate General Relativity we need an invariant volume element and the gravitational equivalent of a gauge covariant derivative. To define an invariant volume element, first note that

$$d^4x \to d^4\tilde{x} \equiv \det\left(\frac{\partial \tilde{x}^{\mu'}}{\partial x^\mu}\right) d^4x. \tag{7.3}$$

If we define

$$-g \equiv \det(g_{\mu\nu}), \tag{7.4}$$

then

$$(-g) \to (-\tilde{g}) = \left[\det\left(\frac{\partial \tilde{x}^{\mu'}}{\partial x^\mu}\right)\right]^{-2} (-g) \tag{7.5}$$

and the invariant volume element is

$$\sqrt{-g}\, d^4x = \sqrt{-\tilde{g}}\, d^4\tilde{x}. \tag{7.6}$$

We next formulate the gravitational covariant derivative. Derivatives of scalar fields transform as

$$\partial_\mu \phi \to \partial_{\mu'} \phi = \frac{\partial x^\mu}{\partial \tilde{x}^{\mu'}} \partial_\mu \phi. \tag{7.7}$$

For a derivative acting on a vector field transforming as $V^\mu \to V^{\mu'} = \frac{\partial \tilde{x}^{\mu'}}{\partial x^\mu} V^\mu$ one finds

$$\partial_\mu V^\nu \to \partial_{\mu'} V^{\nu'} = \left(\frac{\partial x^\mu}{\partial \tilde{x}^{\mu'}} \partial_\mu\right)\left(\frac{\partial \tilde{x}^{\nu'}}{\partial x^\nu} V^\nu\right)$$

$$= \frac{\partial x^\mu}{\partial \tilde{x}^{\mu'}} \frac{\partial \tilde{x}^{\nu'}}{\partial x^\nu} (\partial_\mu V^\nu) + \frac{\partial x^\mu}{\partial \tilde{x}^{\mu'}} \frac{\partial^2 \tilde{x}^{\nu'}}{\partial x^\nu \partial x^\mu} V^\nu. \qquad (7.8)$$

To avoid and cancel the second term on the right hand side of Eq. (7.8) we need to replace the simple derivative ∂_μ by the gravitational covariant derivative

$$\nabla_\mu V^\nu = \partial_\mu V^\nu + \Gamma^\nu_{\mu\lambda} V^\lambda. \qquad (7.9)$$

Here, the $\Gamma^\nu_{\mu\lambda}$ are called connection coefficients. They satisfy the transformation rule

$$\Gamma^\nu_{\mu\lambda} \to \Gamma^{\nu'}_{\mu'\lambda'} = \frac{\partial x^\mu}{\partial \tilde{x}^{\mu'}} \frac{\partial x^\lambda}{\partial \tilde{x}^{\lambda'}} \frac{\partial \tilde{x}^{\nu'}}{\partial x^\nu} \Gamma^\nu_{\mu\lambda} - \frac{\partial x^\mu}{\partial \tilde{x}^{\mu'}} \frac{\partial x^\lambda}{\partial \tilde{x}^{\lambda'}} \frac{\partial^2 \tilde{x}^{\nu'}}{\partial x^\mu \partial x^\lambda}. \qquad (7.10)$$

Equation (7.9) is the gravitational analogue of the gauge covariant derivatives in particle physics. The connection coefficients have a natural expression in terms of the metric and its derivatives

$$\Gamma^\sigma_{\mu\nu} = \frac{1}{2} g^{\sigma\rho}\left(\partial_\mu g_{\nu\rho} + \partial_\nu g_{\rho\mu} - \partial_\rho g_{\mu\nu}\right) \qquad (7.11)$$

known as Christoffel symbols. For covariant derivatives of tensors with lower indices, there is an additional minus sign and a change of the index which is summed over: $\nabla_\mu V_\nu = \partial_\mu V_\nu - \Gamma^\lambda_{\mu\nu} V_\lambda$. One can check that the covariant derivative of the metric vanishes

$$\nabla_\sigma g_{\mu\nu} = \nabla_\sigma g^{\mu\nu} = 0. \qquad (7.12)$$

Information about the curvature of a spacetime manifold is encoded in the Riemann curvature tensor $R^\sigma_{\ \mu\alpha\beta}$ which is defined like the field tensor in gauge theories,

$$(\nabla_\mu \nabla_\nu - \nabla_\nu \nabla_\mu)V_\rho \equiv -R^\lambda_{\ \rho\mu\nu} V_\lambda, \qquad (7.13)$$

where

$$R^\sigma_{\ \mu\alpha\beta} \equiv \partial_\alpha \Gamma^\sigma_{\mu\beta} - \partial_\beta \Gamma^\sigma_{\mu\alpha} + \Gamma^\sigma_{\alpha\lambda}\Gamma^\lambda_{\mu\beta} - \Gamma^\sigma_{\beta\lambda}\Gamma^\lambda_{\mu\alpha}. \qquad (7.14)$$

All components of $R^{\sigma}{}_{\mu\alpha\beta}$ vanish if and only if the (four-dimensional) spacetime is flat (meaning that there is a global coordinate system in which the metric components are everywhere constant). One can define two useful contractions: the Ricci tensor

$$R_{\alpha\beta} = R^{\lambda}{}_{\alpha\lambda\beta} \qquad (7.15)$$

and Ricci scalar

$$R = g^{\mu\nu} R_{\mu\nu}. \qquad (7.16)$$

Using symmetry properties under exchange of indices of the Riemann curvature tensor one finds the Bianchi identity

$$\nabla_{[\lambda} R_{\mu\nu]\rho\sigma} = 0, \qquad (7.17)$$

where we anti-symmetrise over the indices λ, μ and ν. We next define the Einstein tensor

$$G_{\mu\nu} \equiv R_{\mu\nu} - \frac{1}{2} R g_{\mu\nu}. \qquad (7.18)$$

Then, applying the Bianchi identity, one obtains

$$\nabla^{\mu} G_{\mu\nu} = 0, \qquad (7.19)$$

so that the Einstein tensor is gravitation covariantly conserved.

In the presence of matter we would like to couple gravity to the matter source. This requires a symmetric tensor. Einstein's guess was to take the energy momentum tensor $T_{\mu\nu}$ which is then also covariantly conserved, viz.

$$\nabla^{\mu} T_{\mu\nu} = 0. \qquad (7.20)$$

One thus obtains Einstein's equations of General Relativity

$$R_{\mu\nu} - \frac{1}{2} g_{\mu\nu} R = -\frac{8\pi G}{c^4} T_{\mu\nu} + \Lambda g_{\mu\nu}. \qquad (7.21)$$

Here, Λ is the cosmological constant. It is a number and is independent of the local metric $g_{\mu\nu}$.

Einstein's equations link the geometry of spacetime to the energy-momentum tensor. The metric in the equations of motion for General Relativity determines the geodesics — that is, how particles move under the influence of the gravitational field. The cosmological constant term $\Lambda g_{\mu\nu}$ appears both as an integration constant in the covariant conservation of the Einstein tensor (following from Eq. (7.12)) and as an invariant in the Einstein–Hilbert gravitational action

$$S = \int d^4x\sqrt{-g}\left(\frac{1}{2\kappa}(R + 2\Lambda) + \mathcal{L}_M\right). \qquad (7.22)$$

Here, $\kappa = 8\pi G/c^4$ and $\mathcal{L}_M$ is the Lagrangian term describing the matter and radiation contributions to $T_{\mu\nu}$.

When we consider the Standard Model coupled to gravitation $T_{\mu\nu}$ refers to the energy-momentum tensor for excitations above the vacuum. The cosmological constant Λ is interpreted (up to a numerical factor) as the vacuum energy density perceived by gravitation, see Chapter 9. The cosmological constant is the same at all points in spacetime and drives the accelerating expansion of the Universe, see Chapters 8 and 9. The matter and radiation contributions described by $T_{\mu\nu}$ clump according to the laws of usual gravitational attraction.

Newton's theory of gravitation is recovered as the low curvature limit of General Relativity for slowly moving particles. Consider a metric which is almost Minkowski, but with a specific kind of small perturbation $ds^2 = (1 + 2\Phi/c^2)c^2dt^2 - (1 - 2\Phi/c^2)d\vec{x}^2$ where $\Phi = -GM/r$. The 00 component of Einstein's equations is just the Poisson equation for Newtonian gravity $\nabla^2\Phi = 4\pi G\rho$, with ρ the mass density of the gravitating matter.

The SEP is a step beyond the EEP and WEP. With the SEP gravitation proceeds through minimal coupling via the spacetime connection with no extra direct matter to curvature coupling. If present, the latter might change the dynamics without changing the coordinate transformation symmetries. If General Relativity is treated as a low energy effective theory, such direct matter-to-curvature coupling terms might enter as higher dimensional terms in the action, suppressed by powers of the characteristic energy for General Relativity treated as an effective theory (or the scale of emergence if General Relativity might be emergent). These terms would have the effect of changing the geodesics relative to pure General Relativity, with paths in principle different for different particle species. Experimental constraints are discussed in Chapter 11. Any spacetime dependence of G might enter through treating G not as a fixed number but through the matrix element of a new scalar field, the so called scalar-tensor Brans-Dicke theories as an extension of Einsteinian General Relativity [Brans and Dicke (1961)]. (Extended theories of gravitation, beyond minimal General Relativity are reviewed in [Capozziello and De Laurentis (2011)].) General Relativity is thought to be the only theory of gravitation satisfying the SEP.

Unlike the Standard Model of particle physics, General Relativity is a classical theory. If it should be quantised, quantum effects will enter just at very high energies of order the Planck mass[1]:

$$M_{\text{Pl}} = \sqrt{\frac{\hbar c}{G}} \approx 1.2 \times 10^{19} \text{ GeV}. \qquad (7.23)$$

Gravitational forces are negligible in laboratory based particle physics scattering experiments. This follows from comparing the size of the QED fine structure constant $\alpha \approx 1/137 \approx 0.0073$ with its gravitational analogue

$$\alpha_G = GM_{\text{P}}^2/\hbar c = \frac{M_{\text{P}}^2}{M_{\text{Pl}}^2} \approx 5.9 \times 10^{-39}, \qquad (7.24)$$

where M_{P} is the proton mass.

7.2   Tests of General Relativity

General Relativity has proved very successful everywhere the theory has been tested from laboratory tests of the equivalence principle through to large distance effects including gravitational lensing, black holes and new gravitational waves observations.

7.2.1   *Equivalence principle*

The most accurate test of the WEP presently comes from the MICRO-SCOPE experiment in space [Touboul *et al.* (2022)], where it was found to be working at $\mathcal{O}(10^{-15})$, improving on laboratory torsion balance pendulum type experiments [Wagner *et al.* (2012)] which have precision $\mathcal{O}(10^{-13})$ and atom interferometer measurements with precision $\mathcal{O}(10^{-12})$ [Asenbaum *et al.* (2020)].

Atomic and nuclear clocks are used to make precision tests of the EEP. With the EEP dimensionless quantities such as α and the ratio of the electron to proton mass, μ_{ep}, should be velocity and spacetime position independent. Any time dependence in these quantities should show up in atomic and nuclear spectra, including clock transition frequencies. If interaction with dark matter might change fundamental constants this

---

[1]Numerically, Newton's constant G, the speed of light c and Planck's constant $\hbar$ are

$$G = 6.67430(15) \times 10^{-11} \text{m}^3\text{kg}^{-1}\text{s}^{-2} = 6.70883(15) \times 10^{-39}\hbar c \text{ (GeV}/c)^{-2}$$

$$c = 299\ 792\ 458 \text{ ms}^{-1}$$

$$\hbar = 1.054\ 571\ 817\ldots \times 10^{-34} \text{ Js} = 6.582\ldots \times 10^{-22} \text{ MeVs}.$$

would also change the rate at which a clock ticks. Clock measurements are reviewed in [Safronova (2019)]. Prime atomic systems are Al, Sr and Yb clocks. A recent precision measurement [Lange *et al.* (2021)] of possible time dependence in "slow drift" (LPI test) measurements based on the E2 and E3 transitions of $^{171}\mathrm{Yb}^{+}$ gives $\dot{\alpha}/\alpha = (1.0 \pm 1.1) \times 10^{-18}$ year^{-1} and $\dot{\mu}_{ep}/\mu_{ep} = (8 \pm 36) \times 10^{-18}$ year^{-1}. Clock experiments test possible spacetime dependence today. They complement astrophysics constraints on possible variations between now and the early Universe [Hart and Chluba (2018); Ubachs (2018)]. Studies of the Cosmic Microwave Background give $\alpha/\alpha_0 = 0.9993 \pm 0.0025$, where α_0 is the fine structure constant today and α is its value at the time of recombination when the Universe was 380,000 years old [Hart and Chluba (2018)]. Measurements of molecular clouds in space give $|\Delta\mu_{ep}/\mu_{ep}| < 5 \times 10^{-6}$ for redshift $z = 2.0$–4.2, corresponding to look-back times of 10–12.5 Gyrs, and $|\Delta\mu_{ep}/\mu_{ep}| < 1.5 \times 10^{-7}$ for $z = 0.88$, corresponding to half the present age of the Universe (both at 3σ statistical significance) [Ubachs (2018)]. Developments with new quantum sensing technologies will push the frontier with precision tests of the WEP and EEP [Bass and Doser (2024); Ye and Zoller (2024)].

For the SEP experimentally Newton's Law has been shown to work down to 52 μm in recent torsion balance pendulum experiments [Lee *et al.* (2020)] without the need for any modification of General Relativity from possible extra fifth force scalar interactions. This length is less than the dark-energy length scale $\lambda = (\hbar c/\rho_{\mathrm{vac}})^{\frac{1}{4}} \sim 85\mu$m corresponding to the vacuum energy density $\rho_{\mathrm{vac}} \sim 3.8$ keV/cm^3 — or $(0.002$ eV$)^4$ — extracted from astrophysics experiments. Possible time dependence of G is presently constrained from a range of experiments from the solar system to the Cosmic Microwave Background, CMB, with the measured bound typically about $\dot{G}/G < 10^{-11}$ year^{-1} [Chiba (2011)]. Further experiments with low-energy neutrons are looking for evidence of particles associated with possible extra fifth forces with constraints discussed in [Sponar *et al.* (2021)].

For theoretical discussion of possible time dependent fundamental constants including masses and couplings see [Calmet and Keller (2015); Fritzsch (2010); Uzan (2011)].

7.2.2   *Double pulsars and gravitational radiation*

The Double Pulsar system PSR J0737-3039A/B provides a precise test of General Relativity. This is the first known binary system containing two neutron stars both emitting regular radio pulses. It is also the most relativistic binary pulsar discovered so far with higher mean orbital velocities and accelerations than those of other binary pulsars. Precision tests of General Relativity observables in the strong-field regime have now verified the theory at the 0.01% level [Kramer *et al.* (2006, 2021)] with the physics involving emission of gravitational waves. The double pulsar measurements provide indirect evidence for gravitational waves. Direct detection measurements are discussed in Section 7.4.

7.2.3 *Gravitational lensing*

The bending of light by gravity can lead to gravitational lensing whereby multiple images of the same distant astronomical object are visible in the sky. A large scale precision test of General Relativity involves determination of a quantity E_G that combines measurements of large-scale gravitational lensing, galaxy clustering and the growth rate of structure. Measurements are, in general, in good agreement with the predictions of General Relativity with values of E_G obtained to a precision of 5.8–10.7% [Grimm *et al.* (2024)]. Gravitational lensing phenomena is one key piece of evidence for the existence of dark matter.

7.3 Black Holes

Black holes are formed when the gravitational forces due to a compact mass are so strong that the escape velocity exceeds the speed of light. They are regions of space from which nothing, not even light, can escape. Black holes are a prediction of General Relativity [Chandrasekhar (1985); Penrose (1969)] and are now well established by observations [Genzel *et al.* (2024)].

Mathematically, the simplest case is the idealised static Schwarzschild black hole solution. A gravitating body with mass $\mathcal{M}$ becomes a black hole when this mass is contained within the Schwarzschild radius

$$r_s = \frac{2G\mathcal{M}}{c^2} \tag{7.25}$$

with physics described by the Schwarzschild metric

$$ds^2 = \left(1 - \frac{2G\mathcal{M}}{rc^2}\right)dt^2 - \frac{dr^2}{1 - \frac{2G\mathcal{M}}{rc^2}} - r^2(d\theta^2 + \sin^2\theta d\phi^2). \tag{7.26}$$

Infalling particles cross the horizon at the Schwarzschild radius, which is not a real singularity, and then lose communication with the outside world.

There is a mathematical singularity at the centre where the curvature becomes infinite signalling the limit of classical physics. When one goes through the horizon there is a change in the signs of the first two terms in Eq. (7.26) corresponding to an interchange of space and time. The radius becomes timelike while time becomes spacelike. Inside the horizon the "future" points towards $r = 0$ instead of $t = \infty$. The extension from static Schwarzschild to rotating Kerr black holes as well as to charged Reissner–Nordström and rotating plus charged Kerr–Newman black holes is discussed in [Chandrasekhar (1985)]. These come with more complicated internal horizon and singularities structures. Black holes appear to an external observer described just by their mass, spin and charge with all "information" about infalling material hidden behind by the horizon. That is, they have "no hair".

In astrophysics black holes are the endpoint of stellar collapse for stars that are too massive to form neutron stars. There are several mass limits for stars when they collapse at the end of their life cycle with their final state depending on the initial stellar mass. The first is the Chandrasekhar limit. Stars with masses less than 1.39 solar masses $M_\odot$ will end as a white dwarf with the stellar collapse limited by the Pauli principle for electrons. For heavier stars the Landau–(Tolman)–Oppenheimer–Volkoff limit tells us that collapse ends with a neutron star when the mass of the collapsing object is slightly below 3 $M_\odot$ (or 15–20 $M_\odot$ for the original stellar mass before the supernova blows much of this original mass off). Neutron stars are the most dense known QCD matter. For yet heavier stars, the collapse results in a black hole since then the gravitational attraction exceeds the Pauli repulsion from highly dense neutrons in the neutron star interior and there is nothing to stop further collapse. These black holes can be inferred by tracking the movement of a group of stars that orbit a region in space. Alternatively, when gas falls into a stellar black hole from a companion star, the gas spirals inward, heating to very high temperatures and emitting large amounts of radiation that can be detected in experiments.

Supermassive black holes with masses from 10^5 up to a few billion solar masses are observed to occupy the centres of galaxies. The Milky Way contains a supermassive black hole at its centre called Sagittarius A* with mass about 4 million $M_\odot$. A recent highlight in gravitational physics is the observation and imaging of supermassive black hole horizons. The first supermassive black hole imaging focused on the black hole at the centre of the galaxy M87 [Akiyama *et al.* (2019a, 2019b)]. More recently, these measurements have been extended to the Sagittarius A* supermassive black hole at the centre of the Milky Way [Akiyama *et al.* (2022)].

The dynamical origin of these supermassive black holes is an active topic in gravitation research with ideas discussed in [Volonteri *et al.* (2021)]. There are interesting phenomenological correlations between the masses of supermassive black holes at galaxy centres and the masses of their host galaxies [Ferrarese *et al.* (2006)] including their dark matter components [Ferrarese (2002); Ferrarese and Ford (2005)]. The masses of supermassive black holes at galaxy centres observed so far in galaxies at low redshifts are observed to be correlated with the mass of the galaxy dark matter halos as

$$\frac{M_\bullet}{10^8 M_\odot} \sim 0.1 \left(\frac{M_{DM}}{10^{12} M_\odot} \right)^{1.65}. \tag{7.27}$$

Here, $M_\bullet$ is the mass of the supermassive black hole, $M_\odot$ is the solar mass and M_{DM} is the mass of the dark matter halo. This relation is likely to be connected to the dynamics of galaxy formation. In general, the mass of the central massive object at the centre of galaxies is about 0.18% of the galaxy mass with one standard deviation error bars of 0.06–0.52%. Supermassive black holes are dominant in galaxies with total mass bigger than about $10^{10} M_\odot$ [Ferrarese *et al.* (2006)].

Besides supermassive black hole observations, key developments in the studies of black holes include the observation of black hole merger events in gravitational waves detection experiments (see below). Of special interest, new gravitational waves measurements have revealed evidence for intermediate mass black holes that are believed to be too heavy to have formed from a single stellar collapse. Theoretically, there are also ideas involving black holes that might have formed in the early Universe before the time of stellar origin including possible primordial black holes created in the very early Universe.

Black holes also come with an entropy and a black body temperature known as the Hawking temperature T_H [Bekenstein (1973); Hawking (1974, 1975)]. For a Schwarzschild black hole this is

$$T_H = \frac{\hbar c^3}{8\pi G \mathcal{M} k_B} = \frac{\hbar c}{4\pi r_s k_B}, \tag{7.28}$$

where $k_B = 8.617 \times 10^{-5}$ eV K^{-1} is Boltzmann's constant. When the Hawking temperature T_H is above the temperature of space, e.g. the temperature of the CMB 2.726 K, then the black hole behaves as a thermodynamic black body with accompanying radiation. The black hole "evaporates" through so called Hawking radiation into photons and other particles within the allowed mass parameter space. The key process involves quantum pair production at the horizon. One particle falls into the black hole and the other escapes to "infinity". The endpoint of this Hawking radiation is not known and is connected to the fate of the net quantum information that is lost to external observers behind the horizon. It is related to the issue of the black hole horizon meeting the singularity at the endpoint of black hole evaporation. The Hawking temperature for a black hole formed from stellar collapse is about $T_H < 10^{-7}$ K. The lifetime of a stellar mass black hole is about 10^{57} times the present age of the Universe. The black hole lifetime scales as the third power of the black hole mass, $\mathcal{M}_{BH}^3$, and so is very much larger for supermassive black holes. Hawking radiation becomes an issue for possible primordial microscopic black holes that might have formed in the very Universe. If these should form an important part of the dark matter that we observe in astrophysics today, then the original primordial black hole masses cannot have been so small that they would have already evaporated, see e.g. [Green and Kavanagh (2021)].

7.4   Gravitational Waves

An exciting recent development is the discovery of gravitational waves. Predicted by Einstein [Einstein (1915, 1918)], gravitational waves involve the propagation of disturbances in spacetime. Consider the simple case of Einstein's equations in the absence of matter and any cosmological constant contribution,

$$R_{\mu\nu} = 0. \tag{7.29}$$

Writing any fluctuation in the metric as

$$g_{\mu\nu} = \eta_{\mu\nu} + h_{\mu\nu}, \tag{7.30}$$

where $\eta_{\mu\nu}$ is the Minkowski metric contribution, one finds that $h_{\mu\nu}$ satisfies the wave equation

$$\Box \, h_{\mu\nu} = 0. \tag{7.31}$$

The physics of gravitational waves with interactions of massive bodies is described in the textbooks [Maggiore (2007, 2018)]. Gravitational waves involve propagation of disturbances in spacetime with signals which can be triggered during cataclysmic events involving stars or mergers of black holes or neutron stars.

Strong initial indirect evidence for gravitational waves came from the binary pulsar observations discussed above. For merger events involving two massive bodies each with masses $\mathcal{M}_{[1,2]}$, e.g. black holes or neutron stars, the frequency of gravitational waves emitted by two coalescing massive objects is two times the frequency of the motion

$$f = \frac{1}{\pi} \sqrt{\frac{G\mathcal{M}}{r^3}} \qquad (7.32)$$

with $\mathcal{M} = \mathcal{M}_1 + \mathcal{M}_2$. The binary systems are nearly Keplerian. The black holes cannot get closer than their horizons. Substituting the black hole Schwarzschild radius for r gives

$$f \ll \frac{1}{\pi\sqrt{8}} \frac{c^3}{G\mathcal{M}} \qquad (7.33)$$

far from the black holes. Mergers involving supermassive black holes with much larger masses will involve lower frequency gravitational wave emission than for stellar black holes. Typical physics processes that can be observed at different gravitational wave frequencies together with the related detection experiments are shown in Fig. 7.1.

The present gravitational waves detectors are Virgo near Pisa in Italy, the Laser Interferometer Gravitational-Wave Observatory (LIGO) in the United States plus the Kamioka Gravitational Wave Detector (KAGRA) experiment in Japan. These experiments use few kilometer length Michelson interferometers to detect the gravitational waves. They are sensitive to gravitational waves with frequencies in the range 10–10^3 Hz. For experimental details and the experimental discovery of gravitational waves see the Nobel lectures [Barish (2018); Thorne (2018); Weiss (2018)].

The initial discovery gravitational event GW150914 was measured by the LIGO/Virgo collaborations. This involved the merger of two black holes with masses $35.6^{+4.8}_{-3.0} \, M_\odot$ and $30.6^{+3.0}_{-4.4} \, M_\odot$ to yield a remnant black hole with mass $63.1^{+3.3}_{-3.0} \, M_\odot$. The mass difference of $3.1 \pm 0.4 \, M_\odot$ was radiated away through gravitational waves [Abbott *et al.* (2016a, 2016b)]. This black hole merger event occurred at a distance of 430^{+150}_{-170} Mpc. The peak emission rate in the final few milliseconds was $\approx 4 \times 10^{49}$ Watts, more than the combined power of all light emitting stars in the observable Universe. In general, black hole mergers are expected to result in $\approx 5\%$ of the total mass of the black hole system being radiated through gravitational waves with the exact amount being between 3% and 10% depending on the relative spin axes [Barausse *et al.* (2012)].

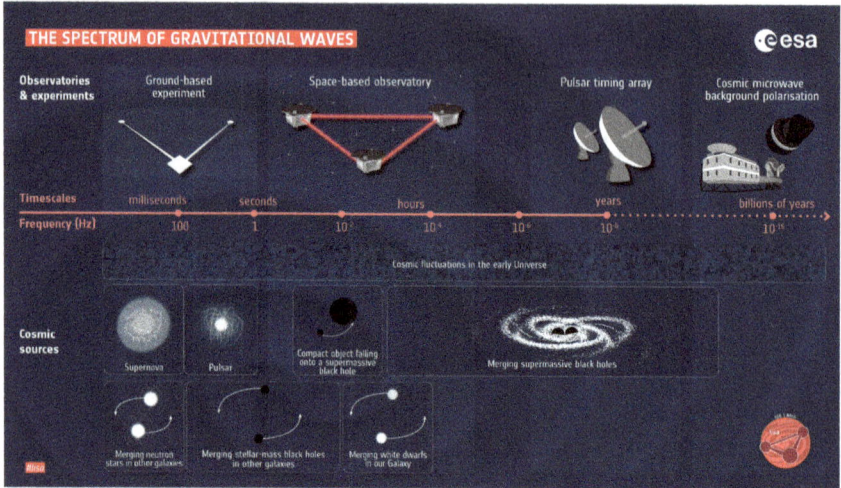

Fig. 7.1. The gravitational wave spectrum showing the phenomena that appear in measurements at different gravitational wave frequencies and the detectors used in these measurements. Figure credit: ESA.

Following the initial gravitational waves discovery, numerous two black hole, black hole — neutron star and two neutron star merger events have been detected. The initial observation involved the merger of two approximately equal mass black holes with the black hole masses in the range expected from stellar collapse. Since then, asymmetric mass black hole mergers and mergers involving more massive black holes have been observed. The event GW190412 was an asymmetric merger of two black holes with masses $29.7^{+5.0}_{-5.3}$ $M_\odot$ and $8.4^{+1.8}_{-1.0}$ $M_\odot$ at distance of 730^{+150}_{-170} Mpc with the remnant black hole having mass $37.0^{+4.1}_{-3.90}$ $M_\odot$. For this event, the experiments were able to determine that the larger black hole was spinning [Abbott *et al.* (2020a)]. The event GW190521 is the heaviest black hole merger seen so far and involved black holes with masses 85^{+21}_{-14} $M_\odot$ and 66^{+17}_{-18} $M_\odot$ at a distance of 5300^{+2400}_{-2600} Mpc leaving a remnant black hole with 142^{+28}_{-16} $M_\odot$ [Abbott *et al.* (2020b)]. This event is especially interesting since one does not expect black holes with masses between about 65 and 120 $M_\odot$ from stellar collapse processes. Intermediate mass black holes might be formed from the mergers of smaller black holes or from some new processes in the early Universe.

Neutron star merger events also provide essential information on the neutron star equation of state [Abbott *et al.* (2017a, 2017b)] inspiring new theoretical work on the QCD structure of dense matter — for reviews see [Baiotti (2019); Yunes *et al.* (2022)]. They also play a key role in telling us about the origin of heavy elements in the Universe through r-process nucleosynthesis and inspiring new laboratory experiments in nuclear physics [Siegel (2022)].

At much lower frequencies, there are also ideas that the stochastic gravitational wave background suggested by the NANOGrav pulsar array experiment [Agazie *et al.* (2023)] might be connected either to supermassive black holes at the centres of galaxies or to processes at work in the early Universe [Caprini (2024)].

In theories of primordial inflation the polarisation of the CMB becomes sensitive to the details of inflation. A background of gravitational waves produced in inflation should induce so-called B-modes in the CMB and induce a finite tensor-to-scalar ratio that is sensitive to the scale of inflation [Lyth (1997)]. There is a vigorous program to search for these modes and to measure the CMB polarisation [Dunkley (2015); Kamionkowski and Kovetz (2016); Komatsu (2022); Seljak and Zaldarriaga (1997)].

A Roadmap for the development of gravitational waves detection experiments in the 2020s–2030s is given in [Bailes *et al.* (2021)] with key physics opportunities also discussed in [Domcke (2024)]. Besides phenomena of major interest in astrophysics, the next generation of gravitational waves experiments will also provide a new probe for exploration of some of the most challenging questions in particle physics.

The Laser Interferometer Space Antenna (LISA), mission of ESA planned to launch in 2035 will use a triangular laser interferometer array of satellites 3 million km apart to measure the gravitational waves spectrum for frequencies between 10^{-4} Hz and 1 Hz. This experiment will study mergers involving supermassive black holes as well as searching for black holes that might have formed in the early Universe with reach up to redshifts $z < 6$ (or when the Universe was about 900 million years old). It will also be sensitive to effects of any (first order) phase transitions involving physics at the TeV scale complementary to LHC measurements. The LISA physics programme is discussed in [Amaro-Seoane *et al.* (2017); Caprini *et al.* (2016); Colpi *et al.* (2024)].

Other experiments presently under discussion to further cover the gravitational waves frequency spectrum include next generation Earth based laser interferometers (the proposed Einstein Telescope [Maggiore *et al.* (2020)] and Cosmic Explorer [Evans *et al.* (2021)] gravitational wave detectors) as well as new experiments using quantum technologies with cold atom interferometers [Alonso *et al.* (2022); Badurina *et al.* (2021)] plus high frequency gravitational wave detectors [Aggarwal *et al.* (2025)]. These proposed experiments should also look for possible modifications of General Relativity due to any graviton mass or Lorentz violation, the origins of supermassive black holes, possible phase transitions in the early Universe, signals of possible ultralight dark matter particles and cosmic string topological effects in space.

7.5   Quantum Gravity?

It is an open question whether gravity should be quantised or not. If so, quantum effects would be expected to be relevant at the energy scale of the Planck mass $M_{\rm Pl} = \sqrt{\frac{\hbar c}{G}}$ or, equivalently, at the Planck length

$$l_{\rm Pl} = \sqrt{\frac{\hbar G}{c^3}}, \tag{7.34}$$

where classical General Relativity would break down. These energy and distance scales are far beyond the reach of present experiments. If gravitation should be quantised, then (corresponding to usual minimal General Relativity) the messenger particle would be a massless graviton with spin two [Feynman (1963); Feynman *et al.* (1995)]. Attempts to quantise gravitation come with the theoretical challenge that a quantum generalisation of General Relativity is not perturbatively renormalisable. Interactions in General Relativity are characterised by Newton's constant which has mass dimension equal to -2 in natural units. A theory of quantum gravity is presently lacking despite much theoretical effort as discussed in [Loll *et al.* (2022)]. One interesting idea involves asymptotic safety whereby the quantised version of General Relativity might come with a non-perturbative ultraviolet fixed point rendering the quantised theory finite and thus non-perturbatively renormalisable [Percacci (2009); Weinberg (1976)].

If the Standard Model is emergent, then why not also General Relativity as the remaining known "fundamental" gauge theory? If General Relativity might be emergent below the Planck scale, e.g. at the same scale supposed for the Standard Model $\approx 10^{16}$ GeV, then the usual problems of quantising General Relativity would be an artifact of extrapolating through the phase transition that produces it. Gravitons would be long-range collective excitations of more primordial degrees of freedom dissolving in the emergence phase transition below the Planck scale. In the condensed matter system of ^{3}He-A one finds an emergent metric and limiting velocities along with emergent SU(2) and U(1) gauge bosons plus emergent "gravitons". Any emergent gravitation is a step beyond the argument that perturbative renormalisability of an emergent quantum field theory with $J = 1$ excitations below some very large mass scale requires gauge symmetry with the $J = 1$ particles being gauge bosons.

The Particle Data Group quotes limits on the photon and possible graviton masses

$$m_\gamma < 1 \times 10^{-18} \text{ eV}$$
$$m_g < 1.76 \times 10^{-23} \text{ eV}.$$

(7.35)

In multimessenger astronomy light and gravitational wave signals identified with the neutron star merger event GKB170817 were observed to arrive very close to simultaneously constraining the speeds of gravity and light to be the same to within $\sim 10^{-15}$ times the speed of light [Abbott *et al.* (2017a)]. Details are discussed in Chapter 11.

Chapter 8

Cosmology: The Interface of Particle Physics and Gravitation

Precision measurements of the Cosmic Microwave Background and the distribution of galaxies in space through baryon acoustic oscillations reveal a visible Universe that is very close to spatially flat. Cosmology observations also reveal the need for new dark energy and dark matter components in the Universe. The dark energy is responsible for the accelerating expansion of the Universe first seen in observations of type 1a supernovae. The expansion of the Universe has been accelerating in the last five billion years or redshift less than about one with the simplest explanation being a cosmological constant in Einstein's equations. Measurements of galaxy properties, gravitational lensing and the CMB consistently point to the presence of some new non-luminous dark matter beyond the normal baryonic matter made of Standard Model particles. This dark matter is stable in that its component in the energy budget of the Universe is observed to be present equally in measurements of the CMB radiation coming from the early Universe and in measurements of galaxies today.

Open questions include what structure the dark matter might possess, e.g. whether it might be associated with new particles, primordial black holes formed in the early Universe or some new extension of usual gravity theory. How should we understand the dark energy? Can the time dependence of the Universe expansion rate be described using present standard cosmology theory — the Λ-CDM model — and might the dark energy be time (in-)dependent?

8.1 Cosmology and the Expanding Universe

Cosmology models start with the assumptions that the Universe is homogeneous and isotropic. These assumptions are consistent with observations. The metric describing the spacetime of cosmology then takes the general Friedmann—Lemaitre—Robertson—Walker, FLRW, form

$$ds^2 = c^2dt^2 - a^2(t)\left[\frac{dr^2}{1 - kr^2} + r^2(d\theta^2 + \sin^2\theta d\phi^2)\right]. \qquad (8.1)$$

Here the function $a(t)$ is known as the scale factor and tells us the relative sizes of the spatial surfaces. It describes the time evolution of the three-dimensional space. The time t, called cosmic or proper time, is the time measured by a clock at rest in these coordinates. The factor k describes the curvature of three-dimensional space and takes the values $k = 0, +1, -1$ for a spatially flat, closed or open Universe, respectively — see Fig. 8.1.[1]

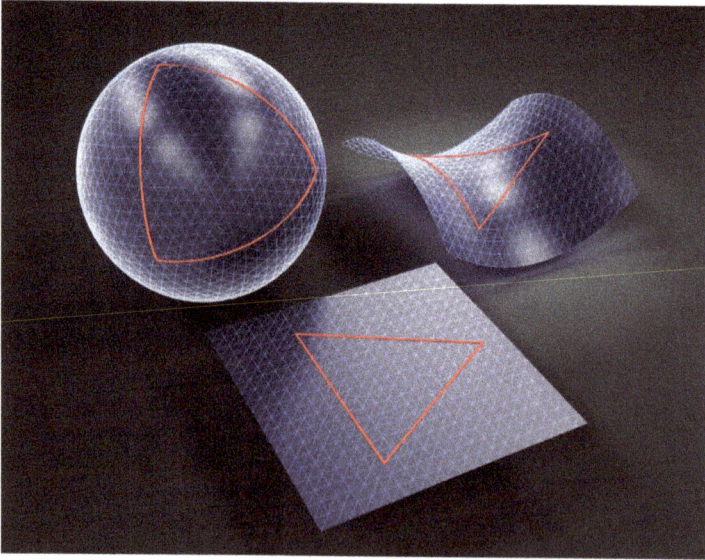

Fig. 8.1. Universe spatial geometry for closed, open and flat Universes with the curvature parameter k equal to $+1$, -1 and 0 respectively. Figure credit: ESO/L. Calcada.

---

[1]For a flat geometry the interior angles of a triangle defined by three beams of light will sum to 180°. For a closed Universe the interior angles of a triangle would sum to more than 180°. For an open Universe the angles would sum to less than 180°.

One can take $a(t)$ carrying units of length, in which case the variable r is unit-less. When $k = \pm 1$, the scale factor $a(t)$ is the radius of curvature of the three-dimensional space. The scale factor a and redshift z are related by

$$1 + z = \frac{a_{\text{now}}}{a_{\text{then}}}, \tag{8.2}$$

where "now" represents the measurement of light today and "then" is the time of its emission.

The Universe is modelled as a perfect fluid characterised by its rest frame energy density ρ and isotropic pressure p. It has no viscosity or shear stresses. The average density ρ and pressure p have the same value everywhere in space but can depend on time. Taking the energy–momentum tensor $T_{\mu\nu}$ as homogeneous and isotropic, we can write it as diagonal in $[\rho, p, p, p]$ through

$$T_{\mu\nu} = (\rho + p)U_\mu U_\nu + p g_{\mu\nu}, \tag{8.3}$$

where U_μ are normalised four-velocities with $g^{\mu\nu}U_\mu U_\nu = -1$.

Solving Einstein's equations (7.21) with the FLRW metric in Eq. (8.1) and this diagonal energy–momentum tensor gives

$$\left(\frac{\dot{a}}{a}\right)^2 = \frac{8\pi G}{3c^2}\rho - \frac{kc^2}{a^2} + \frac{1}{3}\Lambda c^2,$$
$$\frac{\ddot{a}}{a} = -\frac{4\pi}{3c^2}G(\rho + 3p) + \frac{1}{3}\Lambda c^2. \tag{8.4}$$

These equations are known as the Friedmann equations. Here, the Λ cosmological constant contribution has been written explicitly. It is assumed to contain all contributions from the vacuum energy density perceived by gravitation.

The Hubble parameter

$$H = \frac{\dot{a}}{a} \tag{8.5}$$

measures the expansion of the Universe. The ratio of the two Friedmann equations (8.4) defines the deceleration parameter

$$q \equiv -\frac{\ddot{a}a}{\dot{a}^2} = -\left(1 + \frac{\dot{H}}{H^2}\right). \tag{8.6}$$

The expansion of the Universe accelerates if the second derivative of the Universe scale factor, $\ddot{a}$, is positive, which follows if the equation of state of the Universe is such that $w = p/\rho < -1/3$.

The equations of state and solutions for the different types of components (matter, radiation and cosmological constant) when treated in isolation are as follows. One can solve the Friedmann equations (8.4) for possible flat Universes ($k = 0$). For a matter dominated Universe the equation of state has $p = 0$. One finds

$$a(t) \propto t^{\frac{2}{3}},$$
$$\rho_{\text{matter}} \propto a^{-3}. \tag{8.7}$$

For a radiation dominated Universe with just photons the equation of state has $p = \frac{1}{3}\rho$ which corresponds to a traceless energy–momentum tensor. Photons are massless so there is no finite mass scale here. This gives

$$a(t) \propto t^{\frac{1}{2}},$$
$$\rho_{\text{radiation}} \propto a^{-4}. \tag{8.8}$$

For a vacuum dominated Universe the equation of state is $p = -\rho$ (or $T_{\mu\nu} \propto g_{\mu\nu}$ in Eq. (8.3). Here,

$$a(t) \propto e^{Ht},$$
$$\rho_{\text{vac}} = \texttt{constant} = \frac{\Lambda}{8\pi G}, \tag{8.9}$$

with Hubble constant

$$H = \sqrt{\frac{8\pi G \rho_{\text{vac}}}{3}} = \texttt{constant}. \tag{8.10}$$

To understand these results, the expansion of the Universe is the same in all three spatial dimensions. A given volume expands as a^3. When the volume grows the density of matter inside it dilutes and the energy density ρ_{matter} drops inversely with the expanding volume as $\sim a^{-3}$. For photons $\rho_{\text{radiation}}$ drops faster because the wavelength gets stretched and the frequency of radiation falls by an additional factor of $1/a$. The cosmological constant is a property of the vacuum and does not dilute when the Universe expands. If the flat Universe has a finite cosmological constant term it will eventually come to dominate the expansion when the matter and radiation contributions are sufficiently diluted. The result is an eternally accelerating Universe. Going back in time with fixed cosmological constant, we should eventually come to a radiation dominated phase (associated with the Big

Bang). Given the very different time dependences of the various terms, it is interesting that cosmology observations tell us that $\rho_{\text{vac}} \sim 2\rho_{\text{matter}}$ today.

The overall composition of the Universe can be conveniently described by the density parameter Ω which is defined as the average energy density of the Universe divided by the critical density ρ_{crit} needed for a spatially flat Universe. From the first Friedmann equation in Eq. (8.4) we find

$$\Omega \equiv \frac{\rho}{\rho_{\text{crit}}} = \frac{8\pi G \rho}{3c^2 H_0^2} = 1 + \frac{kc^2}{\dot{a}^2}, \tag{8.11}$$

where the critical density

$$\rho_{\text{crit}} = \frac{3c^2 H_0^2}{8\pi G}, \tag{8.12}$$

corresponds to the flat Universe solution and H_0 is the Hubble parameter today.

In general Ω will change with time unless it is equal to one. An open Universe has $\Omega < 1$ and a closed Universe has $\Omega > 1$. From Eq. (8.11), it follows that

$$H^2 = \Omega H^2 - \frac{kc^2}{a^2(t)}, \tag{8.13}$$

which then implies

$$1 - \Omega(t) = -\frac{kc^2}{H^2(t)a^2(t)}. \tag{8.14}$$

If Ω starts positive or negative, then it stays positive or negative with no change of sign in the time evolution. Taking $a(0) \equiv R_0$ defines the curvature radius for $k \neq 0$. If Ω today is very close to one, then the curvature radius is very much greater than the Hubble radius c/H_0 and the effects of curvature are negligible.

The energy density ρ receives contributions from vacuum, radiation and matter contributions

$$\rho = \rho_{\text{vac}} + \rho_{\text{radiation}} + \rho_{\text{matter}}. \tag{8.15}$$

For these three contributions we define $\Omega_i = \rho_i/\rho_{\text{crit}}$. One also defines the curvature contribution $\Omega_k \equiv 1 - \Omega = -kc^2/\dot{a}^2$. For a flat Universe ($k = 0$) the deceleration parameter is

$$q = \frac{1}{2}(\Omega_m + 2\Omega_\gamma + \{1 + 3w\}\Omega_\Lambda), \tag{8.16}$$

where $w = p/\rho$ is the equation of state for the "dark energy" contribution. The solutions of the FLRW equations with a time independent cosmological constant are

$$\left(\frac{H}{H_0}\right)^2 = \Omega_{R,0}(1+z)^4 + \Omega_{M,0}(1+z)^3 + \Omega_{K,0}(1+z)^2 + \Omega_{\Lambda,0}, \quad (8.17)$$

where Ω_R, Ω_M, Ω_K and Ω_Λ are the radiation, matter, curvature and cosmological constant contributions and the subscript 0 denotes their values at the time today.

8.2   Dark Energy and the Λ-CDM Model

Supernovae type 1a, the CMB, Baryon Acoustic Oscillations (ripples in the distribution of galaxies, BAO), and gravitational lensing provide complementary constraints on the fractional energy densities of the matter and dark energy contributions to the energy budget of the Universe. They also tell us about the dark energy equation of state. There is good agreement between measurements from different observables. One finds consistent overlapping confidence level contours that in combination tell us about the contributions of the dark energy and dark matter to the energy budget of the Universe as well as the dark energy equation of state [Frieman *et al.* (2008)].

Traditionally, within the experimental uncertainties, the data have proved consistent with the phenomenological Λ-CDM model, where Λ denotes the (time independent) cosmological constant and CDM denotes cold dark matter. The model uses the FLRW metric and the Friedmann equations, Eq. (8.4), to describe the observable Universe for all times after the initial big bang. The Universe is observed to be very near to spatially flat. Structures like galaxies are understood to have grown out of a primordial linear spectrum of nearly Gaussian and scale invariant energy density perturbations. With just a seven parameters fit to the data the model explains the existence and structure of the CMB, the large scale structure of galaxy clusters and the distribution of light elements (hydrogen, helium, lithium, oxygen) plus the accelerating expansion of the Universe observed in the light from distant galaxies and supernovae [Lahav and Liddle (2024)]. The Λ-CDM model is the simplest model that is, in general, in good agreement with observations though recent tensions have been observed.

The Hubble expansion rate H_0 extracted from the Planck experiment measurements of the CMB which provides a snapshot of the Universe at 380,000 years old is [Aghanim *et al.* (2020)]

$$H_0 = 67.43 \pm 0.49 \text{ kms}^{-1}\text{Mpc}^{-1}. \quad (8.18)$$

This H_0 value corresponds to the critical density

$$\rho_{\text{crit}} = \frac{3c^2 H_0^2}{8\pi G} = 1.05 \times 10^{-5} h_0^2 \text{ GeV cm}^{-3} \qquad (8.19)$$

in units where H_0 is written as $H_0 = h_0 \times 100 \text{ kms}^{-1}\text{Mpc}^{-1}$. The Planck measurement corresponds to $h_0 = 0.674 \pm 0.005$. On the other hand, "recent time" measurements based on distance ladder measurements of low redshift Cepheid variable stars and type 1a supernovae, SNe, by the SH0ES experiment yield a slightly larger value [Riess *et al.* (2022)]

$$H_0 = 73.01 \pm 0.99 \text{ kms}^{-1}\text{Mpc}^{-1} \qquad (8.20)$$

corresponding to $h_0 = 0.730 \pm 0.010$. The interpretation of this "Hubble tension" between these early and late time measurements — a few standard deviation effect — is a major topic of present investigation [Efstathiou (2024); Freedman (2017); Riess (2019); Schöneberg *et al.* (2022); Verde *et al.* (2019)]. Might the Hubble tension be due to something remaining to be understood in the low redshift measurements or evidence for new physics within or beyond the Λ-CDM model? New JWST low redshift measurements combining tip of the Red Giant Branch (TRGB) stars, JAGB (J-Region Asymptotic Giant Branch) stars, and Cepheids give a value

$$H_0 = 69.96 \pm 1.05 \,(\text{stat}) \pm 1.12 (\text{sys}) \text{ kms}^{-1}\text{Mpc}^{-1}, \qquad (8.21)$$

when tied in to SNe data [Freedman *et al.* (2024); Lee *et al.* (2024)]. These JWST numbers are consistent with the Λ-CDM model. The TRGB and JAGB measurements agree at the 1% level and differ from the JWST Cepheid measurements at the 2.5–4% level. It will be interesting to see what consensus develops between the experimental groups. The two values of H_0 in Eqs. (8.18) and (8.20) differ by just 8% after time evolution of the Λ-CDM model between the CMB time and the time today so, at minimum, the Λ-CDM model is providing a good approximation to cosmology observations. A phenomenological look at present tensions in cosmology measurements including the Hubble constant within the Brans–Dicke approach with a time dependent dark energy ("running vacuum") is given in [Solà Peracaula (2022); Solà Peracaula *et al.* (2020)].

Cosmology observations of a spatially flat Universe constrain the energy densities of dark energy and dark matter. The normal baryonic and cold dark matter, plus radiation and dark energy contributions measured by

Planck are [Aghanim *et al.* (2020)]

$$\Omega_b = 0.0493(6), \tag{8.22}$$

$$\Omega_{\text{cdm}} = 0.265(7), \tag{8.23}$$

$$\Omega_\gamma = 2.5 \times 10^{-5}, \tag{8.24}$$

$$\Omega_\Lambda = 0.685(7). \tag{8.25}$$

The total matter contribution is $\Omega_m = \Omega_b + \Omega_{\text{cdm}} = 0.315(7)$, that is the sum over baryonic and cold dark matter contributions. The curvature density parameter comes out as

$$\Omega_k = 1 - \Omega = kc^2/\dot{a}^2 = 0.0007(19), \tag{8.26}$$

with k the curvature and a the scale factor appearing in the FLRW metric suggesting a flat Universe. Observation of a spatially flat Universe together with measurements of the dark matter contribution suggested the presence of a finite cosmological constant [Efstathiou *et al.* (1990)] before the discovery of accelerating expansion of the Universe through supernova Sn1a observations [Perlmutter *et al.* (1999); Riess *et al.* (1998)].

For the dark energy equation of state, EoS, a best fit within the Λ-CDM model gives

$$w = -1.028 \pm 0.031, \tag{8.27}$$

from CMB + BAO + SNe data [Weinberg and White (2024)]. This value is consistent with -1. More general fits can be performed with a more flexible model allowing for possible time dependent dark energy and non-zero curvature. A time dependent dark energy equation of state is frequently parametrised in the form

$$w(a_{\text{then}}) = w_p + w_a \frac{a_p - a_{\text{then}}}{a_{\text{now}}}, \tag{8.28}$$

or, equivalently,

$$w(z) = w_p + w_a \left(\frac{1}{1 + z_p} - \frac{1}{1 + z} \right). \tag{8.29}$$

With this parametrisation one finds $w_p = -1.020 \pm 0.032$ at the pivot redshift $z = 0.29$ with gravitational lensing data also included [Weinberg and White (2024)]. The fit gives a tight constraint on the curvature $1 - \Omega_{\text{tot}} = -0.0023 \pm 0.0032$ but just loose constraint on the evolution parameter $w_a = -0.48^{+0.36}_{-0.30}$. The cosmological constraints on Ω_m and Ω_Λ are illustrated in Fig. 8.2.

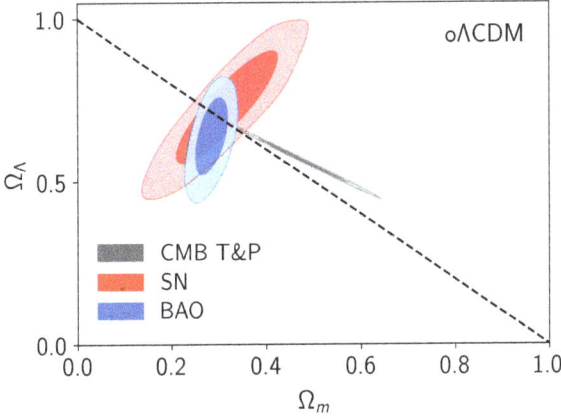

Fig. 8.2. Cosmological constraints under the assumption of a model with a $w = -1$ cosmological constant with free curvature. One finds 68% and 95% constraints on Ω_m-Ω_Λ from the Planck CMB temperature and polarisation data (gray), Pantheon Supernovae, SNe, Ia sample (red), and Sloan Digital Sky Survey BAO-only measurements (blue). The dashed line represents a model with zero curvature. Image credit: the Sloan Digital Sky Survey. For details see [Alam *et al.* (2021)].

Taking $k = 0$ for the flat $w_0 w_a$ model with w_0 corresponding to the pivot value w_p at time today, the time evolution of the Hubble constant behaves as

$$\left(\frac{H}{H_0}\right)^2 = \Omega_{M,0}(1+z)^3 + \Omega_{\Lambda,0}(1+z)^{3(1+w_0+w_a)}e^{-3w_a z/(1+z)}, \quad (8.30)$$

where the radiation component is neglected and the curvature term vanishes for a flat Universe.

8.3 Observations

From ground based studies of supernovae to satellite probes of the CMB, experimental evidence points to the accelerating expansion of the Universe, which was first reported in 1998 [Perlmutter *et al.* (1999); Riess *et al.* (1998)]. This acceleration has occurred in recent cosmic history dating back five billion years corresponding to redshifts of about $z \leq 1$. This is where the acceleration becomes a deceleration due to the lesser impact of ρ_{vac} at earlier times.

Supernova studies of the accelerating Universe use type Ia supernovae. These events are caused by runaway thermonuclear explosions following accretion onto a carbon/oxygen white dwarf star. They have almost uniform brightness making them "standard candles" which can be used for the precise measurement of astronomical distances. Light from these sources is fainter than expected for a given expansion velocity indicating that the supernovae are farther away than predicted with just normal matter densities which in turn indicates that the expansion of the Universe is accelerating.

Measurements of the temperature fluctuations in the CMB provide independent support for the theory of an accelerating Universe. At very early times the temperature was high enough to ionise the material that filled the Universe: the Universe consisted of a plasma of nuclei, electrons and photons, and the number density of free electrons was so high that the mean free path for the Thomson scattering of photons was extremely short. As the Universe expanded, it cooled, and the mean photon energy diminished. Eventually, at a temperature of about 3000 K, the photon energies became too low to keep the Universe ionised. At this time, known as recombination, the primordial plasma coalesced into neutral atoms and the mean free path of the photons increased to roughly the size of the observable Universe. This radiation has since travelled essentially unhindered through the Universe and provides a snapshot of the Universe when it was only 380,000 years old.

The photons we see today in the CMB are a true photograph of the Universe at that time, now 13.7 billion years later. The radiation has cooled to microwave frequencies and is observed as the Cosmic Microwave Background, the thermal afterglow of the Big Bang. To a very good approximation, the temperature of the CMB is uniform across the whole sky. Moreover, it is the most perfect black-body spectrum known with a mean temperature about 2.73 K as first measured by the COsmic Background Explorer (COBE) satellite in 1992 [Smoot *et al.* (1992)]. The discovery of the CMB, together with the black-body nature of its frequency spectrum, was of fundamental importance to cosmology because it validated the idea of a hot Big Bang — the Universe was hot and dense in the past and has since cooled by expansion. Equally important is the fact that the CMB has slight variations of one part in 100,000 in its temperature. The most accurate measurement of these fluctuations was performed by the Planck experiment [Aghanim *et al.* (2020)]. The temperature anisotropies reflect the primordial inhomogeneities in the underlying density field that provided the seeds for cosmological structure formation: galaxies, stars, planets and life. The temperature variations are commonly plotted as a function of the multipole moment — that is, the angular size of the "hot" and "cold" spots. Since the cosmological constant became important only recently in the history of the Universe its main effect is to change the distance to the last scattering surface, which determines the angular size of the CMB anisotropies. Studying the CMB thus yields precise information about the geometry of the Universe — for detailed reviews see [Challinor and Peiris (2009); Durrer (2020)].

Additional evidence for accelerated expansion comes from measured ripples in the distribution of galaxies that were imprinted in acoustic oscillations of the plasma when matter and radiation decoupled as protons and electrons combined to form hydrogen atoms 380,000 years after the Big Bang. These are the "baryonic acoustic oscillations". These probes each indicate a large scale flat Universe geometry.

When the history of the Universe is defined in terms of redshift the CMB corresponds to a snapshot of the Universe at $z \sim 1100$. The Universe changed from radiation to matter dominance at $z \sim 3000$. Most recently, the transition from matter to dark energy dominance happened at $z \sim 0.3$. Originating in the very early Universe there is believed to be an, as yet undetected, cosmic neutrino background corresponding to a snapshot of the Universe at $z \sim 10^{10}$ when neutrinos were first able to propagate freely. Stars were first formed starting when the Universe was about 100 million years old (or $z \sim 30$) with star formation peaking around $z \sim 2$ (or the Universe about 3.3×10^9 years old). Galaxies typically have their origins about 10 to 13.6 billion years ago in the early Universe.[2]

8.4 Dark Matter Hints and Constraints

Some new dark matter (with no electromagnetic interaction) is believed to comprise $\approx 83\%$ of the matter budget of the Universe and 26% of the total energy budget [Baudis (2018); Bertone and Tait (2018); Wechsler and Tinker (2018)]. This dark matter is suggested by studies of galaxies and galaxy clusters, gravitational lensing and the Cosmic Microwave Background. The history of observations and thinking that led to the concept of dark matter is reviewed in [Bertone and Hooper (2018)]. Unlike the cosmological constant or dark energy which is the

---

[2]The age of the Universe as a function of redshift z is given by

$$t(z) = \frac{1}{H_0} \int_0^{1/(1+z)} \frac{dx}{x\sqrt{\Omega_{\Lambda,0} + \Omega_{M,0}x^{-3} + \Omega_{\gamma,0}x^{-4} + \Omega_{K,0}x^{-2}}}. \qquad (8.31)$$

same everywhere, dark matter clumps together under normal gravitational attraction. Whatever the structure of dark matter, it is not comprised of normal baryonic matter. The quest to understand this dark matter has inspired vast experimental and theoretical activity with ideas including new types of elementary particles, primordial black holes as well as possible new theories of gravitation beyond minimal General Relativity and Newtonian dynamics as its low curvature limit.

This non-luminous dark matter is stable or, at least, very long lived with a lifetime greater than the age of the Universe since the freeze-out time of the CMB. If made of particles, the dark matter comes without electric charge. It is cold or not too warm so that the ratio of momentum to particle mass $p/m \ll 1$ at the CMB time. Dark matter is feebly interacting both with itself and with normal baryonic Standard Model matter. The dark matter obeys a matter equation of state.

So far dark matter is known only through its gravitational interaction. The mass range of its possible constituents is presently unknown and spans a vast range of possibilities from possible ultralight new elementary particles up to primordial black holes created in the early Universe. Any new elementary particles might have masses between 10^{-21} eV for new ultralight (pseudo-)scalars up to Planck mass objects. Up to about an eV these dark matter candidates are best described as a classical field. For heavier masses a particle description is used. Dark matter candidates are discussed in Chapter 11.

Chapter 9

Vacuum Energy and the Cosmological Constant

The simplest explanation of the accelerating expansion of the Universe is a cosmological constant Λ in Einstein's equations of General Relativity,

$$R_{\mu\nu} - \frac{1}{2}g_{\mu\nu}\,R = -\frac{8\pi G}{c^4}T_{\mu\nu} + \Lambda g_{\mu\nu}. \tag{9.1}$$

Here, $R_{\mu\nu}$ is the Ricci tensor, R is the Ricci scalar and $T_{\mu\nu}$ is the energy-momentum tensor for excitations above the vacuum; G is Newton's constant and c is the speed of light.[1] Matter including dark matter clumps together under normal gravitational attraction whereas the cosmological constant is the same at all points in spacetime. Einstein's equations determine the geodesics on which particles propagate in curved spacetime in the presence of a gravitational source.

The cosmological constant measures the vacuum energy density perceived by gravitation,

$$\rho_{\text{vac}} = \Lambda \times c^4/(8\pi G), \tag{9.2}$$

with associated scale μ_{vac}, $\rho_{\text{vac}} = \mu_{\text{vac}}^4$. Astrophysics observations [Aghanim et al. (2020)] tell us that $\Lambda = 1.088 \times 10^{-56}$ cm^{-2} corresponding to (in natural units)

$$\rho_{\text{vac}} = (0.002 \text{ eV})^4 \tag{9.3}$$

---

[1] Here, we follow the convention with the Minkowski metric taken as diag$[-1, +1, +1, +1]$ so that positive Λ corresponds to a positive energy density and a negative pressure [Weinberg (1989)].

with a present period of accelerating expansion that began about five billion years ago when the matter density of the expanding Universe fell below ρ_{vac}, which then took over as the main driving term for the expansion.

So far, astrophysics data has proved consistent with dark energy being a time independent cosmological constant with vacuum equation of state (energy density = - pressure) [Escamilla *et al.* (2024)]. Searches for possible time dependence in the dark energy are presently a topic vigorous investigation with new ongoing and planned next generation cosmological surveys. In this chapter, we first discuss a time independent cosmological constant within the framework of an emergent Standard Model and then considerations of how possible time dependence might be included.

The small value 0.002 eV in Eq. (9.3) is intriguing from the viewpoint of Standard Model quantum fields. In general, vacuum energy is sensitive to quantum fluctuations and potentials in the vacuum with terms involving the much larger QCD and electroweak scales (≈ 300 MeV and 246 GeV). If taken alone, these particle physics contributions give large curvature contributions when substituted into Einstein's equations — inconsistent with the flat Universe that we observe in cosmology. This issue for zero-point energies was noticed already in the early works [Pauli (1933); Zeldovich (1967)] with spontaneous symmetry breaking and potentials discussed in [Dreitlein (1974); Veltman (1997)]. (The history of thinking on this topic is reviewed in [Kragh and Overduin (2014); Straumann (2002)].) The net ρ_{vac} corresponding to the cosmological constant also involves an extra gravitational contribution which may be dynamical — for detailed discussion see Section 10.1.

Historically, Einstein introduced the cosmological constant in an attempt to give a static Universe [Einstein (1917)]. Shortly afterwards, he expressed doubts describing Λ as "gravely detrimental to the formal beauty of the theory" [Einstein (1919)]. The static Universe solution proved unstable to local inhomogeneities in the matter density. Einstein abandoned the cosmological constant, setting it equal to zero, following Hubble's observation of an expanding Universe [Einstein (1931)]. Feynman in his lectures on gravitation also wrote that he believed Einstein's second guess and expected a zero cosmological constant [Feynman *et al.* (1995)]. It returned to phenomenology in 1990 with the suggestion that observations of large scale structure imply a spatially flat Universe with finite cosmological constant [Efstathiou *et al.* (1990)], and then in 1998 with discovery of the accelerating expansion of the Universe [Perlmutter *et al.* (1999); Riess *et al.* (1998)].

Vacuum energy only becomes an observable when coupling to gravitation via the cosmological constant. Without gravitational coupling only energy differences count, so in usual particle physics one is free to set the energy of the vacuum to zero, e.g. through normal ordering (before considerations of spontaneous symmetry breaking). The cosmological constant is an observable measurable through the accelerating expansion of the Universe. Hence, it should be particle physics renormalisation scale invariant,

$$\frac{d}{d\mu^2}\rho_{\text{vac}} = 0. \tag{9.4}$$

Here, we take Newton's constant G as renormalisation group scale invariant with gravity treated as classical. Individual contributions are, in general, renormalisation scale dependent (so theorist/calculation dependent) with only the net cosmological constant as the observable. Why is the cosmological constant so small?

The size of the cosmological constant has attracted considerable theoretical attention and ideas. Key historical reviews, each with its own emphasis, include [Bass (2011); Copeland *et al.* (2006); Frieman *et al.* (2008); Martin (2012); Padmanabhan (2003); Peebles and Ratra (2003); Sahni and Starobinsky (2000); Solà (2013); Straumann (2007); Veltman (1997); Weinberg (1989)].

The cosmological constant is connected with the symmetries of the metric. With a finite cosmological constant Einstein's equations have no vacuum solution where $g_{\mu\nu}$ is the constant Minkowski metric. That is, global spacetime translational invariance (a subgroup of the group of general coordinate transformations) of the vacuum is broken by a finite value of ρ_{vac} [Weinberg (1989)].[2] The reason is that a finite value of ρ_{vac} acts as a gravitational source which generates a dynamical spacetime with accelerating expansion for positive ρ_{vac}. Suppose the vacuum including condensates with finite vacuum expectation values is spacetime translational invariant and that flat spacetime is consistent at mass dimension four, just as suggested by the success of the Standard Model. With the Standard Model as an effective theory emerging in the infrared, the low energy global symmetries including spacetime translation invariance can be broken through additional higher dimensional terms, suppressed by powers of the large scale of emergence M.

---

[2]Set $T_{\mu\nu} = 0$ to remove excitations above the vacuum. If the global Minkowski metric is a solution then on the left hand side of Eq. (9.1) there is no curvature. Derivative terms will vanish with $R = 0$ so the cosmological constant will also vanish.

Then the renormalisation group invariant scales Λ_{qcd} and electroweak Λ_{ew} might enter the cosmological constant with the scale of the leading term suppressed by Λ_{ew}/M (that is, $\rho_{\mathrm{vac}} \sim (\Lambda_{\mathrm{ew}}^2/M)^4$ with one factor of $\Lambda_{\mathrm{ew}}^2/M$ for each dimension of spacetime) — see [Bass and Krzysiak (2020a, 2020b)] and the early work [Bjorken (2001a, 2001b)]. This scenario, if manifest in nature, would explain why the cosmological constant scale 0.002 eV is similar to what we expect for the tiny neutrino masses [Altarelli (2005)] which for Majorana neutrinos are themselves linked to the dimension five Weinberg operator with $m_\nu \sim \Lambda_{\mathrm{ew}}^2/M$ [Weinberg (1979)]. That is,

$$\mu_{\mathrm{vac}} \sim m_\nu \sim \Lambda_{\mathrm{ew}}^2/M. \tag{9.5}$$

Here one is taking the Standard Model as describing particle interactions at $D = 4$ up to the large scale M. The cosmological constant would vanish at mass dimension four. This vanishing cosmological constant contribution is equivalent to a renormalisation condition $\rho_{\mathrm{vac}} = 0$ at $D = 4$ being imposed by global spacetime translational invariance of the vacuum even in the presence of QCD and Higgs condensates. The precision of global symmetries in our experiments, e.g. lepton and baryon number conservation, tells us that in this scenario the scale of emergence should be deep in the ultraviolet, much above the Higgs and other Standard Model particle masses. Taking the value $\mu_{\mathrm{vac}} = 0.002$ eV from astrophysics together with $\Lambda_{\mathrm{ew}} = 246$ GeV gives a value for M about 10^{16} GeV.

The scale 10^{16} GeV is within the range where the Higgs self-coupling λ might cross zero (if indeed it does) under renormalisation group evolution. It is also similar to the "GUT scale" that typically appears in unification models. If the Standard Model is emergent at a scale where λ is non-negative then its vacuum will be fully stable. Any perturbative extrapolation of Standard Model degrees of freedom above the scale of emergence would reach into an unphysical region.

Traditionally, present data have been taken as revealing no evidence for a time dependent dark energy. With improved precision the search for possible time dependence as well as any deviations from General Relativity is proceeding with next generation cosmological surveys. There are various ideas how dark energy might come with time dependence. Within the emergence scenario any time dependence might correspond to a change in $\Lambda_{\mathrm{ew}}^2/M$ (with G and c taken as fixed) times its coefficient in the low energy expansion. Here, it is natural to want the EEP to be working at $D = 4$ in Λ_{ew} with any violation of the equivalence principle occurring in the higher dimensional term describing the cosmological constant. In this case, possible time dependence might occur in the scale of emergence or ultraviolet completion M and/or in the coefficient of the Λ_{ew}^2/M term that appears in the low energy expansion. In the first case, any growth in M should not allow it to exceed the Planck scale. Otherwise, quantum gravity effects would enter before the scale of emergence for the Standard Model. Any net time dependence in ρ_{vac} would describe relaxation of ρ_{vac} as one gets time-wise further away from the topological-like phase transition that produces the Standard Model. Condensed matter analogies are described in [Volovik (2005, 2006, 2023)].

As a further example of possible time dependence, de Sitter space is unstable in the S-matrix formulation of quantum gravity with a quantum breaking effect and (slow) decay of the cosmological constant [Dvali (2020); Dvali *et al.* (2017)].[3] (de Sitter space is spacetime with just a cosmological constant and no matter or radiation contributions.) The possible instability of vacuum energy and de Sitter space is also discussed in [Polyakov (2010, 2012)].

Alternatively, possible time dependence might correspond to a vanishing cosmological constant beyond leading order in $1/M$ and with dark energy instead corresponding to some new dynamics with just gravitational coupling, e.g. a BRST invariant coherent state of gravitons as suggested in [Berezhiani *et al.* (2022)] or the time dependent vacuum expectation value of a new ultralight scalar "cosmon" field [Peebles and Ratra (1988, 2003); Wetterich (1988, 1995)] that might interpolate between primordial inflation (an initial burst of accelerating expansion) and cosmology today. This "cosmon" field should have tiny mass $<10^{-33}$ eV or a Compton wavelength bigger than the Hubble length $R_0 \sim 1/H_0$ to explain the uniformity of dark energy in the Universe that we see. In these latter two cases, one then has the phenomenological issue of explaining the similar sizes $\mu_{vac} \sim m_\nu$ observed today. Besides emergence, possible ideas relating the sizes of the cosmological constant scale and neutrino mass are discussed in [Brookfield *et al.* (2006); Fardon *et al.* (2004); Wetterich (2007)].

Cosmology observations of a spatially flat Universe [Aghanim *et al.* (2020)] constrain the energy densities of dark energy and dark matter, see Chapter 8. Observation of a spatially flat Universe together with measurements of the dark matter contribution suggested the presence of

---

[3] If the measured value of the cosmological constant were to saturate the quantum break bound then the classical description of the Universe would work up to 10^{100} years [Dvali *et al.* (2017)], much greater than the present age of the Universe.

a finite cosmological constant [Efstathiou *et al.* (1990)] before the discovery of accelerating expansion of the Universe through supernovae type 1a observations [Perlmutter *et al.* (1999); Riess *et al.* (1998)]. Alternatively, with the cosmological constant scale deduced using the emergent Standard Model arguments above in Eq. (9.5), flatness gives a constraint on the amount of extra dark matter needing new theoretical understanding.

The size of the cosmological constant is also constrained by anthropic arguments. Accelerating expansion of the Universe takes over when the energy density associated with the cosmological constant exceeds the mean matter density (including dark matter contributions). If the cosmological constant were ten times larger the present period of acceleration would have began earlier enough that galaxies would have no time to form [Weinberg (1987)]. With $\rho_{\text{vac}} \sim (\Lambda_{\text{ew}}^2/M)^4$ this constraint corresponds to a factor of 1.33 on Λ_{ew} or upper bound on the Higgs mass through $m_h^2 = 2\lambda v^2$ with v taken as $v = \Lambda_{\text{ew}} = 246$ GeV if we take M as fixed and assume that the dark matter component is not significantly changed by varying Λ_{ew} [Bass and Krzysiak (2020a)].

Besides the present accelerating expansion, it is believed that the Universe experienced an initial period of exponential expansion called inflation, with factor at least 10^{26} in the first about 10^{-33} seconds. The observations of flatness, isotropy in the CMB and homogeneity plus today the lack of magnetic monopoles which might have been produced in the very early Universe motivate primordial inflation ideas where initial exponential expansion is driven by an inflaton scalar field and slow-roll potential condition. For reviews, see [Baumann and Peiris (2009); Ellis and Wands (2024)]. The detailed particle physics of inflation is unknown though there are many theoretical ideas including some where the Higgs boson plays an important role [Bezrukov and Shaposhnikov (2008); Jegerlehner (2014b); Rubio (2019)] or where the vev of a possible time dependent extra scalar "cosmon" field interpolates between initial inflation and dark energy today [Peebles and Ratra (1988, 2003); Wetterich (1988, 1995)]. In this picture the scale of inflation is typically taken as around 10^{16} GeV, which is close to M, our scale of emergence. Perhaps the physics underlying inflation and emergence might be connected.

An interesting experimental probe of inflation ideas involves measurements of the CMB polarisation. In inflation models where usual General Relativity is understood to work between the Planck scale and the scale of inflation, this scale of inflation is related to the tensor-to-scalar perturbation ratio r observable through B modes in the CMB. One finds $V_{\text{inflation}} \sim (\frac{r}{0.01})^{\frac{1}{4}} \times 10^{16}$ GeV [Lyth (1997)]. In these models a finite value of r would be evidence of gravitational waves from the inflationary period. The present experimental constraint is $r_{0.05} < 0.036$ at 95% confidence level from a combination of Planck, WMAP and BICEP/Keck observations [Ade *et al.* (2021)].

While inflation models are phenomenologically very attractive the underlying physics remains an open puzzle. One key issue is the special initial conditions required at the initial singularity in inflation models whcih is connected to entropy and the second law of thermodynamics. The models involve fine tuning of the initial conditions and scalar potential to yield the right amplitude for the density perturbations corresponding to the Universe observed today [Penrose (1989, 2004); Turok (2014)]. It is an open question whether emergence ideas with new degrees of freedom and perhaps new physical laws operating above the scale of emergence might help in resolving this discussion.

Chapter 10

Scale Hierarchies

Particle physics comes with interesting hierarchies of scales. The tiny cosmological constant scale 0.002 eV is much smaller than the electroweak scale 246 GeV which is, in turn, much smaller than the Planck scale 1.2×10^{19} GeV, viz.

$$\mu_{\text{vac}} \ll m_h, \Lambda_{\text{ew}} \ll M_{\text{Pl}}. \tag{10.1}$$

These hierarchies are observed despite the appearance of a quadratically divergent counterterm in the renormalisation of m_h^2 and a quartically divergent counterterm with the zero-point energy contribution to the cosmological constant.

The Higgs boson's observed decays to vector bosons indicate the existence of a Higgs condensate. Its mass $m_h = 125$ GeV was expected to be commensurate with the electroweak scale to ensure the unitarity of the scattering of longitudinally polarised vector bosons. However, such a relatively small mass (relative to the Planck scale that defines the limit of particle physics before quantum gravity effects might appear) raised the fundamental question of the naturalness of the Standard Model. Further, the cosmological constant is tiny when compared to size of the usual particle physics scales that describe quantum fluctuations in the vacuum.

So far we have addressed the cosmological constant within the framework of an emergent Standard Model. In this chapter, we discuss the scale hierarchies in Eq. (10.1) and how they connect with issues of renormalisation and ultraviolet regularisation. We first address the cosmological constant and then the Higgs mass hierarchy puzzle, viz. how to understand $m_h^2 \ll M_{\text{Pl}}^2$.

10.1   Structure of the Cosmological Constant

The cosmological constant through ρ_{vac} receives contributions from the zero point energies, ZPEs, of quantum field theory through vacuum-to-vacuum diagrams [Jaffe (2005)], any (dynamically generated) potential in the vacuum, e.g. induced by spontaneous symmetry breaking and Higgs and QCD condensates [Dreitlein (1974); Veltman (1997)], and the renormalised version of a bare gravitational term ρ_Λ. One finds [Solà (2013); Weinberg (1989)]

$$\rho_{\text{vac}} = \rho_{\text{zpe}} + \rho_{\text{potential}} + \rho_\Lambda. \tag{10.2}$$

As an observable the cosmological constant is renormalisation scale invariant. It is independent of how a theoretician might choose to calculate it,

$$\frac{d}{d\mu^2}\rho_{\text{vac}} = 0 \tag{10.3}$$

with Newton's constant G is taken as renormalisation scale invariant. The individual terms in Eq. (10.2) contributing to ρ_{vac} are, however, renormalisation scale dependent as well as sensitive to large particle physics scales. The Higgs potential is renormalisation group scale dependent through the Higgs self-coupling λ, which determines the stability of the electroweak vacuum. The ZPEs are discussed in detail below. This renormalisation scale dependence should cancel with the gravitational term ρ_Λ to give the scale invariant ρ_{vac}. The important question then is what is left over or how the different terms combine. The small cosmological constant and net ρ_{vac} are determined by the symmetries of the spacetime metric with Nature so much liking the Minkowski metric and a spatially flat Universe today. In any self-consistent discussion, each of the three terms in Eq. (10.2) should be defined with respect to the same renormalisation scheme. The net ρ_{vac} obeys the vacuum equation of state, EoS.

10.1.1   *Zero-point energies*

ZPEs are a vacuum energy contribution induced by quantisation and are an intrinsic part of quantum field theory [Bjorken and Drell (1965)]. One sums over harmonic oscillator contributions. When we evaluate the ZPEs they come with an ultraviolet divergence requiring regularisation and renormalisation.

In quantum physics the ZPEs are not directly measureable. The Casimir effect which is sometimes quoted as evidence for ZPEs can be calculated without recourse to ZPEs and involves Feynman diagrams with external lines whereas the ZPEs involve just closed loops [Jaffe (2005)]. The issue of vacuum energy becomes physical only with coupling to gravity through the cosmological constant.[1]

Working in flat space-time the ZPE for a particle with mass m is

$$\rho_{\text{zpe}} = \frac{1}{2} \sum \{\hbar\omega\} = \frac{1}{2}\hbar \sum_{\text{particles}} g_i \int_0^{k_{\max}} \frac{d^3k}{(2\pi)^3} \sqrt{k^2 + m^2}. \qquad (10.4)$$

Here, m is the particle mass; $g_i = (-1)^{2j}(2j+1)f$ is the degeneracy factor for a particle i of spin j, with $g_i > 0$ for bosons and $g_i < 0$ for fermions. The minus sign follows from the Pauli exclusion principle and the anti-commutator relations for fermions. The factor f is 1 for bosons, 2 for each charged lepton and 6 for each flavour of quark (2 charge factors for the quark and antiquark, each with 3 colours). The corresponding vacuum pressure is [Martin (2012)]

$$p_{\text{zpe}} = \frac{1}{3}\frac{1}{2}\hbar \sum_{\text{particles}} g_i \int_0^{k_{\max}} \frac{d^3k}{(2\pi)^3} \frac{k^2}{\sqrt{k^2 + m^2}}. \qquad (10.5)$$

When evaluating the integrals in Eqs. (10.4) and (10.5) one finds that the equation of state is sensitive to the choice of ultraviolet regularisation. A Lorentz covariant regularisation is needed to ensure that the ZPEs satisfy the vacuum EoS $\rho = -p$.

For example, suppose we evaluate the ZPE using a (non-covariant) brute force cut-off on the divergent integral with some fixed $k_{\max}$. Then the leading term in the ZPE, which is proportional to $k_{\max}^4$, instead obeys the radiation equation of state $\rho = 3p$, viz.

$$\rho_{\text{zpe}}\big|_{k_{\max}} = \hbar\, g_i \frac{k_{\max}^4}{16\pi^2}\left[1 + \frac{m^2}{k_{\max}^2} + \cdots\right],$$

$$p_{\text{zpe}}\big|_{k_{\max}} = \frac{1}{3}\hbar\, g_i \frac{k_{\max}^4}{16\pi^2}\left[1 - \frac{m^2}{k_{\max}^2} + \cdots\right] \qquad (10.6)$$

---

[1]In particle physics before coupling to gravity the ZPEs are commonly "normal ordered" away. More generally, the ZPE may be cancelled by a counterterm consistent with the symmetries of the theory. Beyond the free-field theory ZPEs, comparable contributions are found when interactions are introduced. The vacuum energy is related to the sum of all vacuum-to-vacuum Feynman diagrams. A counterterm can be introduced to cancel these contributions to any order in perturbation theory.

with quadratic terms behaving as $\rho = -3p$ and just subleading logarithmic terms obeying the vacuum EoS [Martin (2012)].[2] Cosmology observations and local energy conservation require that the net vacuum energy contribution ρ_{vac} obeys the vacuum EoS [Peebles and Ratra (2003)]. While the leading term in Eq. (10.6) behaves like an homogeneous sea of radiation, it is important to recall that ρ_{zpe} comes from calculating vacuum-to-vacuum closed loop diagrams instead of freely propagating photons like which dominated in the early Universe.

The vacuum equation of state $\rho = -p$ is obtained with dimensional regularisation and minimal subtraction, $\overline{MS}$. Here, the ultraviolet divergence is controlled by analytic continuation of the theory including the dimensionality of loop integrals in the complex plane and then taking the limit that $d \to 4$ ['t Hooft and Veltman (1972)]. One finds [Brown (1994); Martin (2012)]

$$\rho_{zpe} = -p_{zpe} = -\hbar \, g_i \, \frac{m^4}{64\pi^2} \left[\frac{2}{\epsilon} + \frac{3}{2} - \gamma - \ln\left(\frac{m^2}{4\pi\mu^2} \right) \right] + \cdots \qquad (10.8)$$

for particles with mass m. In Eq. (10.8), $d = 4 - \epsilon$ is the number of dimensions with the divergence at $d \to 4$ or $\epsilon \to 0$ and $\gamma = 0.57721\ldots$ is Euler's constant. The renormalisation scale μ enters also through the running mass and should decouple from the net ρ_{vac}. It enters the calculation of ρ_{vac} to keep the mass dimension of the cosmological constant fixed in Eq. (9.1) [Brown (1994)].[3] After subtracting out the divergent pole

---

[2]The full expressions can readily be obtained by first making the angular integration and then using the hyperbolic substitution $k = m \sinh u$, viz.

$$\rho_{zpe}\big|_{k_{max}} = \hbar g_i \frac{k_{max}^4}{16\pi^2} \left[\sqrt{1 + \frac{m^2}{k_{max}^2}} \left(1 + \frac{1}{2}\frac{m^2}{k_{max}^2} \right) - \frac{1}{2}\frac{m^4}{k_{max}^4} \ln \frac{k_{max}}{m} \left(1 + \sqrt{1 + \frac{m^2}{k_{max}^2}} \right) \right]$$

and

$$p_{zpe}\big|_{k_{max}} = \frac{1}{3}\hbar g_i \frac{k_{max}^4}{16\pi^2} \left[\sqrt{1 + \frac{m^2}{k_{max}^2}} \left(1 - \frac{3}{2}\frac{m^2}{k_{max}^2} \right) + \frac{3}{2}\frac{m^4}{k_{max}^4} \ln \frac{k_{max}}{m} \left(1 + \sqrt{1 + \frac{m^2}{k_{max}^2}} \right) \right].$$

$$(10.7)$$

[3]The full expression for the ZPE term in Eq. (10.4) after covariant dimensional regularisation is

$$\rho_{vac} = -p_{vac} = -\hbar \, g_i \, \frac{\mu^4}{2(4\pi)^{(d-1)/2}} \frac{\Gamma(-d/2)}{\Gamma(-1/2)} \left(\frac{m}{\mu} \right)^d \qquad (10.9)$$

with Eq. (10.8) obtained in the limit $\epsilon \to 0$. Equation (10.9) respects the correct vacuum equation of state. The pole terms "thrown away" by the analytic dimensional continuation come with residue proportional to μ^4 for the pole at $1/(4 - \epsilon)$ and $m^2\mu^2$ for the pole at $1/(2 - \epsilon)$.

term into a renormalisation counterterm, the remaining finite part with $\overline{\text{MS}}$ renormalisation is $(\rho_{\text{zpe}} = -p_{\text{zpe}})|_{\overline{\text{MS}}} = -\hbar \, g_i \, \frac{m^4}{64\pi^2} \left[\frac{3}{2} - \ln\left(\frac{m^2}{4\pi\mu^2}\right) \right]$.

The ZPE in Eq. (10.8) for each given particle is proportional to the particle's mass to the fourth power, thus vanishing for massless photons and gluons. For the Standard Model particles (quarks, charged leptons, W, Z and Higgs) the non vanishing ZPEs are generated through the BEH mechanism. Majorana neutrinos might get their mass through the Weinberg operator $m_\nu \sim \Lambda_{\text{ew}}^2/M$ with mass connected to the electroweak scale Λ_{ew}. The ZPE would vanish for possible massless gravitons. For the Standard Model Higgs with mass squared $m_h^2 = 2\lambda v^2$, the Higgs boson ZPE in Eq. (10.8) develops an imaginary part if the Higgs self-coupling crosses zero through the $m_h^4 \ln m_h^2$ term in Eq. (10.8) signalling vacuum instability for negative λ, $[\ln \lambda = \ln(-\lambda) - i\pi$ for $\lambda < 0]$, see [Bass and Krzysiak (2020a)].

10.1.2 *Interpreting ρ_Λ*

To understand the interpretation of the gravity term ρ_Λ here, the net ρ_{vac} in Eq. (10.2) and the cosmological constant are the observables, not the individual terms ρ_{zpe}, $\rho_{\text{potential}}$ and ρ_Λ in Eq. (10.2) that appear as intermediate steps in the calculation. The cosmological constant term ρ_{vac} is renormalisation scale invariant so the renormalisation scale dependence of the ZPE and potential terms cancels with the gravity term ρ_Λ, which also restores the symmetries of the low energy vacuum below the scale of emergence.[4] If one does wish to use the non-covariant brute force cut-off to define the ZPEs, then one should compensate with a similar Lorentz violation in the ρ_Λ term to guarantee the correct vacuum EoS for the net ρ_{vac}.

The role of ρ_Λ here has an analogy with the treatment of vacuum energy in quantum liquids in condensed matter physics [Volovik (2005)]. The Gibbs–Duhem relation for a quantum liquid at zero temperature with zero pressure sets the corresponding vacuum energy density to zero. If the condensed matter system is in equilibrium then the Gibbs–Duhem relation,

---

[4]Note the part similarity with the Chern–Simons current and the axial anomaly. The Chern-Simons current restores gauge invariance in the renormalised axial vector current and also carries renormalisation scale dependence. Here, the ρ_Λ term is restoring the global symmetry of spacetime translation invariance of the vacuum. The Chern–Simons current in QCD involves gluon fields that couple to the quark axial-vector vertex beyond simple tree approximation, whereas here ρ_Λ appears as a gravitational term in General Relativity.

which is valid in the thermodynamic limit where the number of particles $N \to \infty$, reads

$$E - TS - \mu N = -pV. \qquad (10.10)$$

Here, E is the energy, T the temperature, S the entropy, μ the chemical potential and p the pressure for the system in a volume V. The equilibrium vacuum at zero temperature satisfies the vacuum equation of state

$$\epsilon_{\text{vac}} \equiv \frac{(E - \mu N)}{V} = -p_{\text{vac}}. \qquad (10.11)$$

If there is no external pressure, then the vacuum energy density for the condensed matter system will vanish without extra fine tuning [Volovik (2006)]. One finds cancellation from quasiparticle ZPE quantum contributions up to the energy scale characterising the scale of emergence for the quantum liquid against the macroscopic degrees of freedom, e.g. atoms ... , from above the cut-off for the low energy theory (the latter playing an analogous role to ρ_Λ). Whatever the initial state of the system, it must relax to the equilibrium thermodynamic state where the vacuum energy is nullified without any fine-tuning. A key difference between an emergent Standard Model and these low temperature condensed matter systems is that in the condensed matter systems we know the degrees of freedom both above and below the scale of emergence whereas Planck scale physics in particle physics, e.g. physics above the scale of emergence, is not directly accessible to experiments.

A specific example of Eq. (10.11) for statistical systems is the Ising model with no external magnetic field which satisfies the same vacuum equation of state as the cosmological constant [Bass (2014)].

The QCD phase or crossover transition in the early Universe which occurred when the Universe was $\approx 10^{-5}$ seconds old involves a change in the free energy [Straumann (2004)]. This need not change the cosmological constant if the change in the quark-gluon ZPE and QCD potential term contributions to ρ_{vac} is compensated by a change in the gravity term ρ_Λ keeping the symmetries of the metric held fixed, at least at leading order in $1/M$. This would correspond to a change in the ultraviolet completion of the effective theory with a change in the degrees of freedom from quarks and gluons to hadrons.

There is a second related issue connected with QCD confinement. In hadron physics the degrees of freedom depend on the resolution. Deep in the ultraviolet one has asymptotic freedom. For massless quarks, the $\overline{\text{MS}}$ scheme ZPE in Eq. (10.8) vanishes. Quark-gluon interactions are chiral symmetric at these scales. In the infrared confinement and dynamical chiral symmetry breaking take over: the degrees of freedom are protons, neutrons, pions, nucleon resonances... If energy conservation held for the ZPE (plus QCD potential terms) alone one would find a constraint condition on the hadron spectrum from summing over hadronic ZPE contributions. In practice, just the net ρ_{vac} summing over ZPE, potential and ρ_Λ terms is the observable through coupling to gravity via the cosmological constant.

10.2 The Higgs Mass

The Higgs boson discovered at CERN in 2012 [Aad *et al.* (2012); Chatrchyan *et al.* (2012)] completes the particle spectrum of the Standard Model. In all experimental tests so far it behaves very Standard Model like [Aad *et al.* (2020); Sirunyan *et al.* (2019)]. With the measured Higgs mass the Standard Model satisfies perturbative unitarity (and is renormalisable) with a vacuum that is stable with positive λ up to very high scales. The small size of the Higgs mass compared to the Planck scale raises interesting questions of the naturalness of the Standard Model.

The Higgs boson's mass is very much less than the Planck scale despite quantum corrections, which naively act to push its mass towards the deep ultraviolet. Theoretically, the renormalised Higgs mass squared comes with the divergent counterterm

$$m^2_{h\ \text{bare}} = m^2_{h\ \text{ren}} + \delta m^2_h \tag{10.12}$$

which relates the renormalised and bare Higgs boson masses where, at leading order, employing a cut-off in the Higgs self-energy diagrams gives the quadratic divergence

$$\delta m^2_h = \frac{K^2}{16\pi^2}\frac{6}{v^2}\left(m^2_h + m^2_Z + 2m^2_W - 4m^2_t\right)$$

$$= \frac{K^2}{16\pi^2}2\left(\frac{9}{4}g^4 + \frac{3}{4}g'^4 + 6\lambda - 6y^2_t\right). \tag{10.13}$$

Here, K is an ultraviolet scale characterising the limit to where the Standard Model should work. The renormalised mass is the mass extracted from experiments, being related by a perturbative expansion to the particle pole mass. For a textbook discussion of bare and renormalised mass parameters see Chapter 4. In Eq. (10.13), we use the Standard Model relations connecting the masses and couplings Eqs. (2.75)–(2.77). Smaller contributions from lighter mass quarks and charged leptons are neglected. Next-to-leading-order corrections are suppressed by factors of

$1/(4\pi)^2$ and also neglected. (Their effect is moderate as discussed in [Jegerlehner (2014c); Jones (2013)].) The hierarchy or naturalness puzzle is why is m_h^2 so much smaller than the Planck mass or any other large physical scale that might characterise new physics effects beyond the Standard Model?

If one instead evaluates the Higgs mass squared using dimensional regularisation, then the quadratic divergence in Eq. (10.13) corresponds to a pole at $d = 2$ [Veltman (1981)]. With analytic continuation in the dimensions this pole is "thrown away" in the formal regularisation procedure and the resulting divergence comes out proportional to the $1/\epsilon$ pole term at $d = 4$ instead of the large ultraviolet scale term K^2. Results are available up to two loop calculations, see [Kniehl *et al.* (2015)]. The leading order term with $\overline{\text{MS}}$ is

$$\delta m_h^2 = m_h^2 \frac{1}{16\pi^2} \frac{1}{\epsilon} \left(3y_t^2 + 6\lambda - \frac{9}{4}g^2 - \frac{3}{4}g'^2 \right) + \cdots \qquad (10.14)$$

Here, the scale factor in δm_h^2 is set by the Standard Model particle masses, which follows on dimensional grounds since these are the only mass scales in the calculation. The different coefficients between Eqs.(10.13) and (10.14) follow since the quadratic divergence in Eq. (10.13) corresponds to a pole at $d = 2$ whereas the coefficient in Eq. (10.14) is the pole at $d = 4$ with dimensional regularisation.

If taken alone without coupling to extra particles, the Standard Model is self-consistent without any large scale hierarchy issue if we work in the $\overline{\text{MS}}$ scheme which involves no explicit large mass scale K [Wells (2009)]. Ultraviolet divergences can consistently be absorbed in renormalisation counterterms. However, usual effective theory arguments suggest that, independent of this, mass scales should be close to the ultraviolet cut-off that defines the limit of the effective theory unless some symmetry of the Lagrangian like gauge invariance or chiral symmetry pushes them to zero. The Higgs mass and cosmological constant are not protected this way. Hence, the issue whether and what new physics one needs to suppress them.[5]

---

[5]Note that the ZPEs and Higgs boson mass renormalisation are related through the Standard Model effective potential [Einhorn and Jones (1992); Masina and Quiros (2013)]. (This potential is discussed in detail in e.g. [Pokorski (2000); Taylor (1979)]). Consistency means that they should be addressed within a common regularisation and renormalisation scheme.

In the emergence scenario discussed here it is the dynamics of the vacuum, its stability and spacetime symmetries with an emergent Standard Model that fix the size of the Higgs mass and the cosmological constant with the infrared and ultraviolet connected through renormalisation group running of the Higgs self coupling λ. Vacuum stability involves a delicate fine tuning and conspiracy of Standard Model parameters.

This emergence scenario treats the Standard Model as saturating the particle spectrum at $D = 4$. Another approach to the naturalness puzzle is to assume that the Standard Model quantum correction to the Higgs boson's mass in Eq. (10.13), which is dominated by the top quark contribution, might be cancelled by any new particles that couple to the Higgs boson. However, such particles have so far not been seen in the mass range of the LHC. Likewise, any composite structure to the Higgs boson would soften the ultraviolet divergences, but there is no evidence for this in the present data suggesting any compositeness should be manifest only at energies above the LHC. Searches for extra particles and a possible composite Higgs structure will continue in the coming years with the increased luminosity at the LHC.[6]

In thinking about the Higgs mass hierarchy puzzle with electroweak symmetry breaking, $v \ll M_{\mathrm{Pl}}$, it is interesting also to consider possible similar phenomena with the ferromagnetic phase transition in condensed matter physics. Below the phase transition the magnetisation is very small close to the phase transition when we approach the critical temperature T_c from below, viz. when the reduced temperature $(T - T_c)/T \to 0^-$. Whereas emergent gauge systems are associated with topological like phase transitions without a local order parameter, Higgs phenomena is associated with spontaneous symmetry breaking defined with respect to a particular gauge choice. Electroweak symmetry breaking might correspond

---

[6]Theoretical attempts to resolve the naturalness puzzle include weakly coupled models with a popular candidate being supersymmetry [Wess and Zumino (1974)], which if present in Nature would be a new symmetry between bosons and fermions. Strongly-coupled models, where the Higgs boson is considered as a bound state of new dynamics strong at the weak scale, are an alternative solution to SUSY. In such scenarios the "lightness" of the Higgs boson can be explained if the Higgs boson turns out to be a pseudo-Nambu–Goldstone boson. Such models include the so-called little Higgs [Arkani-Hamed *et al.* (2001, 2002)], twin Higgs [Chacko *et al.* (2006)] and partial compositeness [Kaplan (1991)] models. For a comprehensive review of these ideas and their phenomenology see [Carena *et al.* (2024)], with possible alternatives to an elementary Higgs boson also discussed in [Csaki *et al.* (2016)].

to a Universe close to the phase transition and very near to the critical point [Jegerlehner (2014c)].

It is interesting also to consider the theoretical issue whether the coefficient in Eq. (10.13), might cross zero, viz. at

$$2m_W^2 + m_Z^2 + m_h^2 = 4m_t^2 \qquad (10.15)$$

with collective cancellation between bosons and fermions, and then change sign. Taking the PDG pole masses for the W, Z and top quark masses (80, 91 and 173 GeV) would require a Higgs mass of 314 GeV, much above the measured value. In practice, the running couplings in Eqs. (2.75)–(2.77) have different renormalisation group behaviour, see Fig. 3.1, and the different contributions enter with different signs for bosons and fermions. With renormalisation group scale dependent couplings the result is calculation dependent. If δm_h^2 were to cross zero — so called Veltman crossing — below the Planck scale and below the scale of emergence, then this might be interpreted in terms of a first order phase transition with electroweak symmetry breaking below the crossing scale. Above this scale the Standard Model would enter a symmetric phase with the 4 Higgs states associated with the BEH Higgs doublet field then carrying large mass of order the crossing scale. All the other Standard Model particles would be massless. This scenario was found in the calculations of [Jegerlehner (2014c)] (with emergence taken at the Planck scale instead of 10^{16} GeV) with a stable vacuum up to the Planck mass and Veltman crossing around 10^{16} GeV. If manifest in Nature, it would allow the Higgs to behave as the inflaton [Jegerlehner (2014b)]. In other calculations crossing was found not below the Planck mass in [Bass and Krzysiak (2020a)] and (with metastable vacuum) around 10^{20} GeV in [Masina and Quiros (2013)] and much above the Planck scale in [Degrassi *et al.* (2012)] and [Hamada *et al.* (2013)].

For illustration Fig. 10.1 shows the renormalisation group running of the coefficient

$$C_{V1} = \frac{3}{v^2}\left(m_h^2 + m_Z^2 + 2m_W^2 - 4m_t^2\right) = \frac{9}{4}g^4 + \frac{3}{4}g'^4 + 6\lambda - 6y_t^2 \qquad (10.16)$$

in the Higgs mass counterterm, Eq. (10.13), evaluated with the Standard Model evolution code [Kniehl *et al.* (2016)] that was used in Figs. 3.1 and 3.2. Here, Veltman crossing is found at the Planck scale with a Higgs mass about 150 GeV and not below with the measured mass of 125 GeV. If one takes input values $m_t = 171$ GeV and $m_h = 125$ GeV in this calculation corresponding to a stable vacuum then, in this calculation, Veltman crossing happens not below the Planck scale.

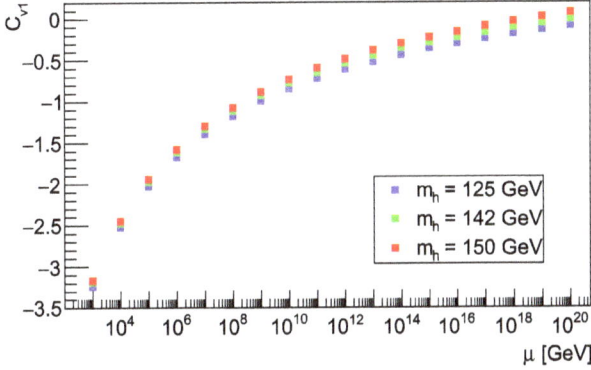

Fig. 10.1.   Running of the Veltman coefficient, Eq. (10.16) for Standard Model particles. Here $C_{V1} = \frac{3}{v^2}(m_h^2 + m_Z^2 + 2m_W^2 - 4m_t^2) = \frac{9}{4}g^4 + \frac{3}{4}g'^4 + 6\lambda - 6y_t^2$ is evaluated using the running couplings in Fig. 3.1. The points are for Higgs masses m_h equal to 150, 142 and 125 GeV (top to below). The figure is taken from [Bass and Krzysiak (2020a)]. In an alternative calculation [Jegerlehner (2014c)], the $m_h = 125$ GeV curve crosses zero at $\mu \approx 10^{16}$ GeV.

The extension to models with possible extra Higgs states is discussed in [Bass and Krzysiak (2020a)].

Similar discussion can be extended to the renormalisation group evolution of the net ZPE contribution. In early work Pauli suggested a collective cancellation of the ZPE contributions [Pauli (1971)] (see also [Kamenshchik *et al.* (2018); Visser (2019)]), much like the Veltman condition for the Higgs mass squared. Figure 10.2 shows the renormalisation group evolution of the net ZPE contribution summed over top, W, Z and Higgs contributions and evaluated in the $\overline{\text{MS}}$ scheme using Eq. (10.8). One finds contributions proportional to

$$\mathcal{P}_1 = 6m_W^4 + 3m_Z^4 + m_h^4 - 12m_t^4$$
$$\mathcal{P}_2 = 6m_W^4 \ln m_W^2 + 3m_Z^4 \ln m_Z^2 + m_h^4 \ln m_h^2 - 12m_t^4 \ln m_t^2. \tag{10.17}$$

In the figure, the renormalisation group running of these two Pauli constraint terms is shown for the case $m_h = 125$ GeV and $m_t = 173$ GeV. This calculation again uses the evolution code in [Kniehl *et al.* (2016)]) with the masses first expressed in terms of Standard Model couplings using Eqs. (2.75)–(2.77). Here, the first Pauli condition $\mathcal{P}_1$ with terms $\propto m^4$ crosses zero above 10^{16} GeV corresponding to the net bosonic ZPE contribution outgrowing the fermionic top quark contribution. The second Pauli condition $\mathcal{P}_2$ is shown up to 10^{10} GeV, above which λ becomes

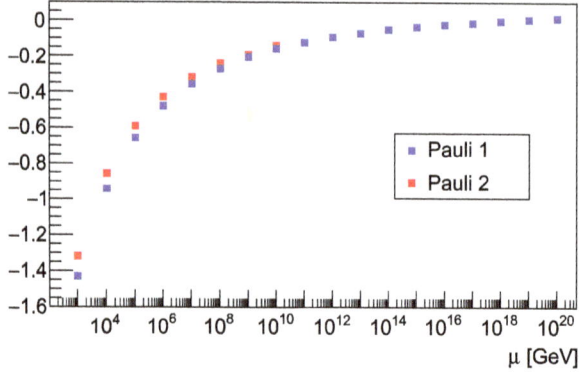

Fig. 10.2.   Running values of the Pauli conditions in Eq. (10.17) summed over Standard Model particles (bosons–fermions), e.g. Pauli 1: $\{6m_W^4 + 3m_Z^4 + m_h^4 - 12m_t^4\}$ and Pauli 2: $\{6m_W^4 \ln m_W^2 + 3m_Z^4 \ln m_Z^2 + m_h^4 \ln m_h^2 - 12m_t^4 \ln m_t^2\}$. The Pauli 1 (lower) points are normalised to v^4. The Pauli 2 (upper) points are normalised to $v^4 \ln v^2$ and plotted up to the scale that they develop an imaginary part when λ crosses zero just above $\mu = 10^{10}$ GeV signaling vacuum instability. Here, the values $m_h = 125$ GeV and $m_t = 173$ GeV are taken. Figure from [Bass and Krzysiak (2020a)].

negative and the combination $m_h^4 \ln m_h^2$ develops an imaginary part. For the stable vacuum case with inputs $m_t = 171$ GeV and $m_h = 125$ GeV both Pauli curves cross zero between 10^{17} and 10^{18} GeV in these calculations. With a stable vacuum λ remains positive definite so that v remains finite and the second Pauli condition develops no imaginary part.

Chapter 11

Emergence and Open Puzzles

In the emergence picture the Standard Model is treated as an effective theory with characteristic energy about 10^{16} GeV. The Standard Model at $D = 4$ is supplemented by a tower of new higher dimensional operator contributions involving the Standard Model fields plus also couplings to possible new particles entering at $D \geq 5$. These extra terms mean that we become sensitive to new physics with possible new global symmetry breaking terms suppressed by powers of the large scale of emergence. The new terms offer possible explanations of tiny (Majorana) neutrino masses, baryogenesis with new sources of CP violation, and facilitate possible axion dark matter. In a low energy expansion one might also consider possible matter to curvature coupling as an SEP violating extension to General Relativity. Beyond the particle degrees of freedom of the Standard Model and possible light new particle extensions like axions, an extra consideration involves what might happen with fluctuations of the system above the scale of emergence which get frozen out if the Standard Model was "born" in some topological-like phase transition in the early Universe. Such fluctuations might be akin to lattice vibrations as an extension to the Fermi–Hubbard model giving long-range bosonic phonon excitations in addition to the local gauge fields.

11.1 Neutrino Masses

Majorana neutrinos can enter at $D = 5$ through the Weinberg operator with accompanying lepton number violation. These neutrino mass terms

$$m_\nu \sim \Lambda_{\mathrm{ew}}^2/M \qquad (11.1)$$

can come with flavour mixing as well as (up to three) possible CP violating complex phases.

11.2   Baryogenesis

The matter–antimatter asymmetry is one of the key open problems at the interface of particle physics and cosmology. Left over from the early Universe, the number of baryons compared to antibaryons is finite and exceeds the number of photons in the Universe by a factor [Navas *et al.* (2024)]

$$\eta_B = (n_B - n_{\bar{B}})/n_\gamma \sim 6 \times 10^{-10}. \qquad (11.2)$$

Sakharov identified three famous conditions which must be satisfied in any theory which claims to explain the observed baryon asymmetry of the Universe [Sakharov (1967)]:

(1) Baryon number non-conservation.
(2) CP and C violation so that processes occur at different rates for particles and antiparticles.
(3) Departure from thermal equilibrium. Otherwise if the Universe starts with zero baryon number it will stay with zero baryon number.

How might higher dimensional operators help? The CP violation in the Standard Model at $D = 4$ is encoded through the CKM matrix. This CP violation is not sufficient to allow for baryogenesis. Extra CP beyond the Standard Model CKM matrix may be provided via possible complex phases with Majorana neutrino masses entering at $D = 5$. One also finds new $D = 6(+)$ operators which violate CP [Grzadkowski *et al.* (2010)]. Also, the $D = 5$ Pauli term could generate a net contribution to the CP violating electron EDM. If present, these terms can become important in the early Universe at energies close to the scale of emergence. Baryon number violation can enter at $D = 6$ [Weinberg (1979); Wilczek and Zee (1979)] in addition to lepton number violation at $D = 5$ [Weinberg (1979)]. Together with non-equilibrium at the phase transition yielding emergence one finds at least qualitatively the conditions needed for baryogenesis. The details of understanding how these effects might combine and whether the effect is large enough to explain the matter–antimatter asymmetry in the Universe is a challenge for theory.

This picture is different to other main ideas for understanding the matter–antimatter asymmetry with a possible first order phase transition involving the Higgs vev generated in the early Universe [Morrissey and Ramsey-Musolf (2012); Trodden (1999)]. Bubbles with a Higgs condensate would be created and expand at the speed of light. This scenario also requires new sources of CP violation beyond the usual Standard Model. Ideas are discussed in [Servant (2018)] and also involve an extended Higgs sector (for example, including an extra singlet scalar) and quantum tunneling processes in the vacuum called sphalerons that violate baryon number conservation. An extended Higgs sector is a step beyond the simplest emergence scenario with the minimal number of particle species and multiplets at $D = 4$. Evidence for any first order electroweak phase transition in the early Universe might show up in future gravitational wave measurements.

11.3 Dark Matter Candidates

Some extra stable and non-luminous dark matter without electromagnetic coupling is suggested by studies of galaxies and galaxy clusters, gravitational lensing and the Cosmic Microwave Background; for reviews see [Baudis (2018); Bertone and Tait (2018); Wechsler and Tinker (2018)]. In cosmology observations we see just a 5% contribution to the energy budget of the Universe from visible matter; 68% is in dark energy and 27% is in dark matter observed (so far) just through gravitational effects. The quest to understand this dark matter, $\approx 84\%$ of the total matter budget, has inspired vast experimental and theoretical activity with ideas including possible new types of elementary particles, primordial black holes as well as possible extensions of General Relativity.

We know that dark matter must be stable by comparing the results of early Universe CMB data and the dark matter contribution today measured through astrophysics observations, with time evolution of the Universe described using the ΛCDM "standard model" of cosmology.

The mass scale of possible new dark matter particles is so far not known from experiments with ideas including new particles with masses ranging from 10^{-22} eV up to 10^{15} GeV — a range of 10^{46} [Baudis (2018)]. Each new particle suggestion comes with its own symmetry properties with different dark matter particle candidates entering at different orders in a low energy expansion. If the Standard Model does describe particle physics interactions at $D = 4$ up to the highest scales, then we do not expect lightest mass supersymmetry particles or extra inert Higgs doublet states as sources of dark matter. With emergence small multiplets are preferred as collective excitations of the system that resides deep in the ultraviolet [Jegerlehner (2014c)]. One might then look to terms suppressed by powers of $1/M$, whether new particle couplings or modified gravity scenarios beyond minimal General Relativity. If present, the pseudoscalar axions discussed in Chapter 5 are a new particle candidate. They would enter at $D = 5$ with masses and couplings to Standard Model particles suppressed by a single power of the large axion "decay constant" [Kawasaki and Nakayama (2013)] which might be associated with any scale of emergence.[1] Also discussed in the literature are very heavy Wimpzilla candidates [Kolb *et al.* (1999)]. These are possible particles with mass $\approx M$ — that is, close to the GUT scale or the ultraviolet scale of completion for an emergent Standard Model.

Besides possible new particles, primordial black holes are discussed as candidates for dark matter [Carr and Kuhnel (2022); Green (2024); Green and Kavanagh (2021)] and are consistent with General Relativity with couplings up to $D = 4$. These suggestions include the black holes in the mass range observed with the LIGO–Virgo gravitational wave measurements [Bird *et al.* (2016); Boehm *et al.* (2021)]. The limits on black hole masses that could saturate the dark matter contribution is presently a prime topic of investigation. Particular interest is focused also on asteroid mass black holes [Green (2024)]. Primordial black holes formed in the early Universe before the time of stellar collapse origin might be detected in future gravity waves measurements [Bertone *et al.* (2020)].

In the context of emergence one might also consider localised perturbations involving degrees of freedom that occur above the scale of emergence. Analogies with phenomena in condensed matter physics are discussed in [Jegerlehner (2023); Klinkhamer and Volovik (2017)]. For example, perturbations in the physics associated with ρ_Λ might have occurred in the early Universe and couple to gravitation with a dark matter equation of state. Dark matter might perhaps behave similar to collective vibrations of the atomic ion lattice in superconductors. Consider the extension of

---

[1]QCD axion masses are usually taken from astrophysics and cosmology constraints between 1 μeV and 3 meV corresponding to an ultraviolet scale between 6×10^9 and 6×10^{12} GeV. This is less than the 10^{16} GeV discussed here in connection with the cosmological constant and neutrino masses. Possible relaxation of the axion bounds to include perhaps higher scales is discussed in [Dvali (1995)]. Any axion Bose–Einstein condensate contribution would approximately scale as $D = 4$ with the axion decay constant suppression factor missing in the net condensate contribution [Sikivie and Yang (2009)].

the Fermi–Hubbard model discussed in Chapter 6 to include collective fluctuations of the atomic lattice sites. The strongly correlated electron system described by the Fermi–Hubbard model develops emergent gauge symmetries in the low energy limit at half filling with the atomic lattice kept fixed. Vibrations of the atomic lattice system (phonons) exhibit bosonic statistics independent of the fermionic or bosonic nature of the atoms making up the lattice structure [Moessner and Moore (2021)]. As a second example, the main aspects of the BCS theory of low temperature superconductivity involve a two-fluid model where the compressible electron gas and the related phonon excitations propagate in an incompressible ion background. Two types of collective vibrational fields occur: the phonons that populate the long-range branch and the displacement field of the essentially incompressible ion system which could act as a dark matter like field [Jegerlehner (2023)]. Such ideas need detailed investigation to see whether they might have a chance to work.

11.3.1 *Galaxy properties*

Whilst we search for evidence of possible new dark matter particles and/or black holes formed in the early Universe it is worthwhile also to consider other observables that might give clues to the nature of dark matter. Interesting information may follow from studying the redshift dependence of correlations between dark matter and galaxy properties, and also common trajectories followed by particle signals from distant galaxies. One observes phenomenological correlations between galaxy rotation velocities and dark matter [Wechsler and Tinker (2018)] and also correlations between the masses of supermassive black holes at galaxy centres and their dark matter halos [Ferrarese (2002); Ferrarese and Ford (2005)] as discussed in Chapter 7. The latter correlations have so far been measured just for galaxies at small redshifts. It will be interesting to measure the time evolution of galaxy formation and these correlations. Measurements of early Universe galaxies at the peak of galaxy formation 10 billion years ago suggest that these galaxies may have contained smaller dark matter components as determined from their rotation curves [Genzel *et al.* (2017)].

The baryonic version of of the Tully–Fisher relation connects the total baryonic mass in a galaxy $\mathcal{M}_{bar}$ to the asymptotic rotation velocity v_∞ [McGaugh *et al.* (2000)], viz.

$$v_\infty^4 = G\mathcal{M}_{bar}a_0, \tag{11.3}$$

where a_0 is a phenomenological acceleration parameter. One finds $a_0 \approx 1.2 \times 10^{-10}$ ms^{-2} to within 10–20% from galaxy rotation curves [Milgrom (2020)]. Numerically $2\pi a_0 \approx a_H \approx a_\Lambda$ where $a_H = cH_0 \approx 1.1 \times 10^{-10}$ ms^{-2} and $a_\Lambda = c^2\sqrt{\Lambda/3} = cH_\infty \approx 0.9 \times 10^{-10}$ ms^{-2}. Here, H_0 is the Hubble parameter today, Λ is the cosmological constant and H_∞ is the Hubble constant in the infinite future of the Λ-CDM model scenario after matter and radiation density terms have decayed approaching zero with the cosmological constant taken as time independent. Equation (11.3) is sometimes taken as phenomenological evidence to hint that usual Newtonian gravity might be modified, the so called MOND theories [Milgrom (1983)].[2] It may also be the net result that should come out from simulating a microscopic theory of dark matter involving new dark matter particles within usual General Relativity and connected to galaxy formation. This issue is a step beyond the simple fit to data associated with Eq. (11.3). Here, we just note that a_H and a_Λ have different scaling behaviour in a $1/M$ expansion with $\mu_{\text{vac}} \sim \Lambda_{\text{ew}}^2/M$: a_H has a mass dimension four component whereas $a_\Lambda \sim 1/M^2$. Further, a_H is time dependent through H_0 whereas a_Λ is time independent if we take a time independent cosmological constant. It would be interesting to measure the redshift dependence of the phenomenological relation in Eq. (11.3) in the evolution of galaxy structures from their peak formation time through to today. Might the value of a_0 be time independent or its similarity with the Hubble constant just a coincidence at low redshift? Possible redshift dependence of a_0, including with Λ-CDM dynamics, is discussed in [Mayer *et al.* (2022)]. If a_0 were found to be redshift independent from some time after the initial formation, then it might hint at dark matter being associated with some higher dimensional coupling involving the cosmological constant.

11.3.2   *Common geodesics*

Another interesting constraint is that the gravitational effect of dark matter or any modified gravity scenarios is the same for all particles. Whatever dark matter is, within present experimental uncertainties it does not seem to affect the common trajectories travelled by gravitational waves and by different particle species [Baudis (2018)]. In multimessenger astronomy light

---

[2]These involve an extra asymptotic long-range gravitational acceleration $g = a_0 \frac{r_M}{r}$ with $r_M = (\mathcal{M}G/a_0)^{\frac{1}{2}}$ where $\mathcal{M}$ is the central mass.

and gravitational wave signals identified with the neutron star merger event GKB170817 were observed to arrive very close to simultaneously constraining the speeds of gravity and light to be the same to within $\sim 10^{-15}$ times the speed of light, meaning that they follow the same geodesics [Abbott *et al.* (2017a)]. Likewise, one might consider the initial neutrino multimessenger event IceCube-170922A with energy 290 TeV [Aartsen *et al.* (2018)] and the neutrino event IceCube-141209A with $E_\nu = 97.4 \pm 9.6$ TeV detected in coincidence with the blazar GB6 J1040+0617 in a phase of high γ-ray activity [Kun *et al.* (2023)]. These each correspond to the photon and neutrino speeds being the same to within about 10^{-11} of the speed of light assuming the neutrinos were emitted coincident with the blazar flaring periods.

Of these events, the gravitational wave event GKB170817 was observed in coincidence with a short γ-ray burst, GRB 170817A, at redshift $z = 0.008^{+0.002}_{-0.003}$, with the signals detected 2 seconds apart originating from a neutron star merger more than 10^8 light years away. The neutrino event IceCube-170922A was identified in parallel with very high energy γ-rays from the blazar source TXS 0506+056 at redshift $z = 0.3365 \pm 0.0010$ with the blazar active within about ± 14 days of the 290 TeV neutrino event. The statistically favoured joint blazar excitation and neutrino event IceCube-141209A was observed in a window of high γ-ray activity from the blazar GB6 J1040+0617 at redshift $z = 0.7351 \pm 0.0045$ which was brightest 4.5 days before the neutrino detection.

Different particle trajectories might be induced in exotic modified gravity scenarios with different metrics for gravitational waves coupling to General Relativity without dark matter and for normal matter coupling to General Relativity with dark matter included. Trajectories might also be affected by any new direct matter to curvature couplings which, if present, would involve an extended gravitational action involving a series in $\mathcal{O}(R/M^2)$. Possible higher dimensional terms that might enter here include $1/M^2 R_{\mu\nu} F^{\mu\omega} F^\nu_{\ \omega}$, $1/M^2 R_{\mu\nu} \bar{\psi} D^\mu D^\nu \psi$ and $1/M^2 R_{\mu\nu\rho\sigma} F^{\mu\nu} F^{\rho\sigma}$ with $F_{\mu\nu}$ is the electromagnetic field tensor, ψ the electron field, $R_{\mu\nu}$ the Ricci tensor and $R_{\mu\nu\alpha\beta}$ the curvature tensor. If present, any direct matter-to-curvature coupling terms correspond to non-minimal gravitational coupling beyond just the connection. They would violate the strong equivalence principle, SEP. These terms can be induced by vacuum polarisation in curved spacetime and might be important in the very early Universe when the curvature scale becomes important. Such terms preserve the local symmetries of gravitational interactions but change the dynamics [Charlton *et al.* (2020); Shore (2005)]. They modify the energy momentum tensor and equations of motion and can lead to particles not following geodesics, with the effect in general different for different particles. In addition, as discussed in Chapter 5, possible violations of Lorentz invariance might arise at $\mathcal{O}(\mu_{\text{vac}}/M \sim \Lambda_{\text{ew}}^2/M^2) \approx 10^{-28}$ [Bjorken (2001a)].

Chapter 12

Conclusions

The main surprise from the LHC is that the Standard Model is working so well. Perhaps the Standard Model is more special than previously assumed. If one assumes no coupling to other particles then renormalisation group evolution suggests that the Standard Model with its measured parameters may be working up to very high scales, perhaps up to the scale 10^{16} GeV where Grand Unified Theories have previously been proposed to work or even up to the Planck scale. Crucial here is the renormalisation group behaviour of the Higgs self-coupling λ which is very sensitive to a delicate interplay of "low energy" Standard Model parameters measured in laboratory experiments. Small changes in some of these parameters would lead to a very different theory and to very different physics. The infrared and ultraviolet limits of the Standard Model theory are thus strongly correlated.

The Standard Model is mathematically self consistent but new physics is needed to explain open puzzles with neutrinos, baryogenesis, fermion families, dark energy, dark matter plus the physics of the very early Universe connected with primordial inflation ideas. The answer to these puzzles lies outside the Standard Model truncated to $D = 4$.

If the Standard Model might really work up to large scales of order $\approx 10^{16}$ GeV or more, then a plausible scenario is that it might be emergent with its gauge symmetries "dissolving" in the ultraviolet and the particles as the stable long range excitations of some critical system that resides deep in the ultraviolet and perhaps close to the Planck scale. In this scenario neutrinos would be Majorana particles. New physics most likely resides in higher dimensional operator terms that become important at energies very close to the scale of emergence and which may have been active in the early Universe. The cosmological constant comes out with a scale similar to what we expect for tiny Majorana neutrino masses, $\mu_{\text{vac}} \sim m_\nu \sim \Lambda_{\text{ew}}^2/M$. The question why the cosmological constant is so small becomes the issue of why Nature so likes the Minkowski metric with connection to a spatially flat Universe. Possible time dependent dark energy might be associated with time dependence in the scale of emergence or ultraviolet completion M and/or in the coefficient of the cosmological constant in the low energy expansion in powers of $1/M$ as the Universe relaxes time-wise away from the topological-like phase transition that produced the Standard Model. The Higgs mass would be environmentally selected in connection with vacuum stability, with the hierarchy puzzle a delicate conspiracy of Standard Model parameters in the low energy phase. It is interesting to consider possible connections between the scale of emergence and the scale of inflation. Above the scale of emergence, the physics would be described by new degrees of freedom and perhaps by new physical laws.

Dark matter in this picture, if it is not primordial black holes, might be associated with axion-type particles or with new higher dimensional gravitational couplings beyond minimal General Relativity. It is also plausible that dark matter might be connected with fluctuations of the system that resides above the scale of emergence that became frozen out in the phase transition that produced the Standard Model. As a condensed matter analogy, the Fermi–Hubbard model of strongly correlated electron systems generates new emergent gauge symmetries in the low energy limit ($U \gg t$) at half filling. Additional collective fluctuations of the atomic lattice structure can yield bosonic phonon excitations beyond the electron and gauge boson quasiparticles. These phonon excitations might be analogous to dark matter. Lorentz invariance with an emergent Standard Model might be broken with tiny effects at $D \geq 6$.

Extra CP violation beyond the Standard Model Cabibbo–Kobayashi–Maskawa matrix can arise through complex phases associated with Majorana neutrino masses and through new $D = 6(+)$ operators. Baryon number violation can also occur at $D = 6$ with possible non-equilibrium at the phase transition associated with emergence and electroweak symmetry breaking providing at least qualitatively the conditions needed for baryogenesis [Sakharov (1967)].

The Standard Model effective theory discussed here differs from usual discussions in particle theory [Isidori *et al.* (2024)] where one supposes new interactions at $D = 4$ at higher energies above the range of present experiments. The effect of these interactions is integrated out with redefinition of the low energy parameters as well as being parametrised by higher dimensional terms with mass scale representing the scale of new physics. New heavy particle degrees of freedom would be liberated when one goes through their production thresholds. In the emergence scenario discussed in this book the Standard Model is taken to work at $D = 4$ up to the large scale of emergence. The higher dimensional terms describe new physics from the phase transition that produces the Standard Model. They are unconstrained by requirements of renormalisability at $D = 4$.

With emergence open theoretical questions are the universality class for the Standard Model and any critical dimension for the topological-like phase transition that might produce it. If the Standard Model might be emergent, then what about gravitation? If the gauge symmetries of General Relativity might be emergent at a scale below the Planck mass then this would alleviate the usual problems associated with quantising gravitation. Another interesting question is the origin of fermion families. Lattice regularisation induces fermion doubling issues associated with the discrete spacetime regularisation of the lattice [Creutz (1985); Luscher (2002)]. Perhaps fermion families in the Standard Model might be associated with emergent chiral fermions (relevant to weak interactions) breaking some symmetry of a critical "Planck system" that sits above the scale of emergence. The origin of "quantum" is also an open puzzle with suggestions that quantum physics might itself be emergent [Adler (2004); 't Hooft (2007); Wetterich (2010a, 2010b)]. Ideas here include the possibilty that fermion quantum field theories might be associated at a deeper level with probabilistic cellular automata with classical bits [Wetterich (2021)]. If gravitation and quantum physics might be emergent below the Planck scale, the result might significantly increase the size of any $\mathcal{O}(1/M^2)$ minimum length corrections to Heisenberg's uncertainty relations that could be looked for in the quantum optics experiments discussed in [Pikovski *et al.* (2012)]. For example, taking $M \approx 10^{16}$ GeV gives an enhancement in $1/M^2$ of 10^6 relative to taking $M = M_{\text{Pl}}$.

The similarity of the cosmological scale and values we expect for light neutrino masses is intriguing, as is the similarity between the phenomenological acceleration parameter a_0 for low redshift galaxies and a_Λ involving the cosmological constant. While these may be coincidences, these relations may be hinting at some deeper connection between the infrared and ultraviolet through higher dimensional terms in a low energy expansion in $1/M$ associated with emergence. In the absence of signals of new particles one should at least treat such numerical similarities as possible clues and see where they might lead.

Next generation experiments at the high energy, precision and cosmology frontiers will teach us much about the deeper structure of Nature. These probes will soon be joined by the next generation of gravitational waves experiments which will be sensitive to black holes that might have formed before the time of stellar origin as well as to possible phase transitions in the early Universe and ultralight dark matter candidates. Measurements of the CMB polarisation will tell us about primordial inflation. If the Standard Model does indeed work up to the highest scales, then the vacuum stability might be pointing to deep interconnections between the physics of the infrared and extreme ultraviolet.

Bibliography

Aad, G. *et al.* (2012). Observation of a new particle in the search for the Standard Model Higgs boson with the ATLAS detector at the LHC, *Phys. Lett. B* **716**, pp. 1–29.

Aad, G. *et al.* (2020). Combined measurements of Higgs boson production and decay using up to 80 fb^{-1} of proton-proton collision data at $\sqrt{s} = 13$ TeV collected with the ATLAS experiment, *Phys. Rev. D* **101**, 1, p. 012002.

Aad, G. *et al.* (2021). A search for the dimuon decay of the Standard Model Higgs boson with the ATLAS detector, *Phys. Lett. B* **812**, p. 135980.

Aad, G. *et al.* (2022). A detailed map of Higgs boson interactions by the ATLAS experiment ten years after the discovery, *Nature* **607**, 7917, pp. 52–59, [Erratum: *Nature* **612**, E24 (2022)].

Aad, G. *et al.* (2024a). Characterising the Higgs boson with ATLAS data from Run 2 of the LHC, arXiv:2404.05498 [hep-ex].

Aad, G. *et al.* (2024b). Climbing to the top of the ATLAS 13 TeV data, arXiv:2404.10674 [hep-ex].

Aad, G. *et al.* (2024c). Electroweak, QCD and flavour physics studies with ATLAS data from Run 2 of the LHC, arXiv:2404.06829 [hep-ex].

Aartsen, M. G. *et al.* (2018). Multimessenger observations of a flaring blazar coincident with high-energy neutrino IceCube-170922A, *Science* **361**, 6398, p. eaat1378.

Abbott, B. P. *et al.* (2016a). GW151226: Observation of gravitational waves from a 22-solar-mass binary black hole coalescence, *Phys. Rev. Lett.* **116**, 24, p. 241103.

Abbott, B. P. *et al.* (2016b). Observation of gravitational waves from a binary black hole merger, *Phys. Rev. Lett.* **116**, 6, p. 061102.

Abbott, B. P. *et al.* (2017a). Gravitational waves and gamma-rays from a binary neutron star merger: GW170817 and GRB 170817A, *Astrophys. J. Lett.* **848**, 2, p. L13.

Abbott, B. P. *et al.* (2017b). GW170817: Observation of gravitational waves from a binary neutron star inspiral, *Phys. Rev. Lett.* **119**, 16, p. 161101.

Abbott, R. *et al.* (2020a). GW190412: Observation of a binary-black-hole coalescence with asymmetric masses, *Phys. Rev. D* **102**, 4, p. 043015.

Abbott, R. *et al.* (2020b). GW190521: A binary black hole merger with a total mass of $150 M_\odot$, *Phys. Rev. Lett.* **125**, 10, p. 101102.

Abe, K. *et al.* (2017). Search for proton decay via $p \to e^+ \pi^0$ and $p \to \mu^+ \pi^0$ in 0.31 megaton·years exposure of the Super-Kamiokande water Cherenkov detector, *Phys. Rev. D* **95**, 1, p. 012004.

Abel, C. *et al.* (2017). Search for Axionlike dark matter through nuclear spin precession in electric and magnetic fields, *Phys. Rev. X* **7**, 4, p. 041034.

Ackermann, M. and Helbing, K. (2023). Searches for beyond-standard-model physics with astroparticle physics instruments, *Phil. Trans. Roy. Soc. A* **382**, 2266, p. 20230082.

Ade, P. A. R. *et al.* (2021). Improved constraints on primordial gravitational waves using planck, WMAP, and BICEP/Keck observations through the 2018 observing season, *Phys. Rev. Lett.* **127**, 15, p. 151301.

Adler, S. L. (1969). Axial vector vertex in spinor electrodynamics, *Phys. Rev.* **177**, pp. 2426–2438.

Adler, S. L. (1970). Pertubation theory anomalies, in S. Deser, M. Grisaru, and H. Pendleton (eds.), *Brandeis Lectures on Elementary Particles and Quantum Field Theory* (MIT Press).

Adler, S. L. (2004). *Quantum Theory as an Emergent Phenomenon: The Statistical Mechanics of Matrix Models as the Precursor of Quantum Field Theory* (Cambridge University Press, Cambridge, UK).

Adler, S. L. and Bardeen, W. A. (1969). Absence of higher order corrections in the anomalous axial vector divergence equation, *Phys. Rev.* **182**, pp. 1517–1536.

Adler, S. L. and Boulware, D. G. (1969). Anomalous commutators and the triangle diagram, *Phys. Rev.* **184**, pp. 1740–1744.

Adolph, C. *et al.* (2017). Final COMPASS results on the deuteron spin-dependent structure function g_1^{d} and the Bjorken sum rule, *Phys. Lett. B* **769**, pp. 34–41.

Affleck, I., Zou, Z., Hsu, T., and Anderson, P. W. (1988). SU (2) gauge symmetry of the large-U limit of the Hubbard model, *Phys. Rev. B* **38**, pp. 745–747.

Agazie, G. *et al.* (2023). The NANOGrav 15 yr Data Set: Evidence for a Gravitational-wave Background, *Astrophys. J. Lett.* **951**, 1, p. L8.

Aggarwal, N. *et al.* (2025). Challenges and Opportunities of Gravitational Wave Searches above 10 kHz, arXiv:2501.11723 [gr-qc].

Aghanim, N. *et al.* (2020). Planck 2018 results. VI. Cosmological parameters, *Astron. Astrophys.* **641**, p. A6, [Erratum: *Astron. Astrophys.* **652**, C4 (2021)].

Aidala, C. A., Bass, S. D., Hasch, D., and Mallot, G. K. (2013). The Spin Structure of the Nucleon, *Rev. Mod. Phys.* **85**, pp. 655–691.

Aitchison, I. J. R. and Hey, A. J. G. (2013a). *Gauge Theories in Particle Physics, 40th Anniversary Edition: A Practical Introduction, Volume 1* (Taylor & Francis).

Aitchison, I. J. R. and Hey, A. J. G. (2013b). *Gauge Theories in Particle Physics, 40th Anniversary Edition: A Practical Introduction, Volume 2* (Taylor & Francis).

Aker, M. *et al.* (2024). Direct neutrino-mass measurement based on 259 days of KATRIN data, arXiv:2406.13516 [nucl-ex].

Akiyama, K. *et al.* (2019a). First M87 event horizon telescope results. I. The shadow of the supermassive black hole, *Astrophys. J. Lett.* **875**, p. L1.

Akiyama, K. *et al.* (2019b). First M87 event horizon telescope results. VI. The shadow and mass of the central black hole, *Astrophys. J. Lett.* **875**, 1, p. L6.

Akiyama, K. *et al.* (2022). First Sagittarius A* Event Horizon Telescope results. I. The shadow of the supermassive black hole in the center of the Milky Way, *Astrophys. J. Lett.* **930**, 2, p. L12.

Alam, S. *et al.* (2021). Completed SDSS-IV extended baryon oscillation spectroscopic survey: Cosmological implications from two decades of spectroscopic surveys at the apache point observatory, *Phys. Rev. D* **103**, 8, p. 083533.

Alekhin, S., Djouadi, A., and Moch, S. (2012). The top quark and Higgs boson masses and the stability of the electroweak vacuum, *Phys. Lett. B* **716**, pp. 214–219.

Alonso, I. *et al.* (2022). Cold atoms in space: Community workshop summary and proposed road-map, *EPJ Quant. Technol.* **9**, 1, p. 30.

Altarelli, G. (2005). Neutrino 2004: Concluding talk, *Nucl. Phys. B Proc. Suppl.* **143**, pp. 470–478.

Altarelli, G. (2013a). Collider Physics within the Standard Model: a Primer, arXiv:1303.2842 [hep-ph].

Altarelli, G. (2013b). The Higgs: So simple yet so unnatural, *Phys. Scripta T* **158**, p. 014011.

Altarelli, G. (2014). The Higgs and the excessive success of the standard model, *Frascati Phys. Ser.* **58**, p. 102.

Altarelli, G. and Ross, G. G. (1988). The Anomalous gluon contribution to polarized leptoproduction, *Phys. Lett. B* **212**, pp. 391–396.

Amaro-Seoane, P. *et al.* (2017). Laser Interferometer Space Antenna, arXiv:1702.00786 [astro-ph.IM].

Amati, D. *et al.* (1981). Dynamical gauge bosons from fundamental fermions, *Phys. Lett. B* **102**, pp. 408–412.

Anderson, P. *et al.* (1987). Resonating–valence-bond theory of phase transitions and superconductivity in La_2CuO_4-based compounds, *Phys. Rev. Lett.* **58**, 26, pp. 2790–2793.

Anderson, P. W. (1963). Plasmons, gauge invariance, and mass, *Phys. Rev.* **130**, pp. 439–442.

Anderson, P. W. (1972). More is different, *Science* **177**, 4047, pp. 393–396.

Anderson, P. W. and Brinkman, W. F. (1973). Anisotropic superfluidity in ^{3}He: A possible interpretation of its stability as a spin-fluctuation effect, *Phys. Rev. Lett.* **30**, pp. 1108–1111.

Anderson, P. W. and Morel, P. (1961). Generalized bardeen-cooper-schrieffer states and the proposed low-temperature phase of liquid He^3, *Phys. Rev.* **123**, pp. 1911–1934.

Andreev, V. *et al.* (2018). Improved limit on the electric dipole moment of the electron, *Nature* **562**, 7727, pp. 355–360.

Aoyama, T., Kinoshita, T., and Nio, M. (2018). Revised and improved value of the QED tenth-order electron anomalous magnetic moment, *Phys. Rev. D* **97**, 3, p. 036001.

Arkani-Hamed, N., Cohen, A. G., and Georgi, H. (2001). Electroweak symmetry breaking from dimensional deconstruction, *Phys. Lett. B* **513**, pp. 232–240.

Arkani-Hamed, N. *et al.* (2002). The littlest Higgs, *JHEP* **07**, p. 034.

Arneodo, M. *et al.* (1994). A reevaluation of the Gottfried sum, *Phys. Rev. D* **50**, pp. R1–R3.

Asenbaum, P. *et al.* (2020). Atom-Interferometric test of the equivalence principle at the 10^{-12} level, *Phys. Rev. Lett.* **125**, 19, p. 191101.

Bañuls, M. C. *et al.* (2020). Simulating lattice gauge theories within quantum technologies, *Eur. Phys. J. D* **74**, 8, p. 165.

Badurina, L. *et al.* (2021). Prospective sensitivities of atom interferometers to gravitational waves and ultralight dark matter, *Phil. Trans. A. Math. Phys. Eng. Sci.* **380**, 2216, p. 20210060.

Baikov, P. A., Chetyrkin, K. G., and Kühn, J. H. (2017). Five-Loop running of the QCD coupling constant, *Phys. Rev. Lett.* **118**, 8, p. 082002.

Bailes, M. *et al.* (2021). Gravitational-wave physics and astronomy in the 2020s and 2030s, *Nature Rev. Phys.* **3**, 5, pp. 344–366.

Baiotti, L. (2019). Gravitational waves from neutron star mergers and their relation to the nuclear equation of state, *Prog. Part. Nucl. Phys.* **109**, p. 103714.

Balantekin, A. B. and Kayser, B. (2018). On the properties of neutrinos, *Ann. Rev. Nucl. Part. Sci.* **68**, pp. 313–338.

Banerjee, D. *et al.* (2012). Atomic quantum simulation of dynamical gauge fields coupled to fermionic matter: From string breaking to evolution after a quench, *Phys. Rev. Lett.* **109**, p. 175302.

Banks, T. and Rabinovici, E. (1979). Finite temperature behavior of the lattice abelian Higgs model, *Nucl. Phys. B* **160**, pp. 349–379.

Barausse, E., Morozova, V., and Rezzolla, L. (2012). On the mass radiated by coalescing black-hole binaries, *Astrophys. J.* **758**, p. 63, [Erratum: *Astrophys. J.* **786**, 76 (2014)].

Barceló, C. *et al.* (2016). From physical symmetries to emergent gauge symmetries, *JHEP* **10**, p. 084.

Barceló, C. *et al.* (2021). Emergent gauge symmetries: Yang-Mills theory, *Phys. Rev. D* **104**, 2, p. 025017.

Barish, B. C. (2018). Nobel lecture: LIGO and gravitational waves II, *Rev. Mod. Phys.* **90**, 4, p. 040502.

Baskaran, G. and Anderson, P. W. (1988). Gauge theory of high temperature superconductors and strongly correlated fermi systems, *Phys. Rev. B* **37**, pp. 580–583.

Baskaran, G., Zou, Z., and Anderson, P. W. (1987). The resonating valence bond state and high T_c superconductivity — A mean field theory, *Solid State Comm.* **63**, pp. 973–976.

Bass, S. D. (2004). Anomalous commutators and electroweak baryogenesis, *Phys. Lett. B* **590**, pp. 115–119.

Bass, S. D. (2005). The spin structure of the proton, *Rev. Mod. Phys.* **77**, pp. 1257–1302.

Bass, S. D. (2011). The cosmological constant puzzle, *J. Phys. G* **38**, p. 043201.

Bass, S. D. (2014). The cosmological constant puzzle: Vacuum energies from QCD to dark energy, *Acta Phys. Polon. B* **45**, pp. 1269–1279.

Bass, S. D. (2020). Emergent gauge symmetries and particle physics, *Prog. Part. Nucl. Phys.* **113**, p. 103756.

Bass, S. D. (2021). Emergent gauge symmetries: Making symmetry as well as breaking it, *Phil. Trans. Roy. Soc. A* **380**, 2216, p. 20210059.

Bass, S. D. (2023). The cosmological constant and scale hierarchies with emergent gauge symmetries, *Phil. Trans. Roy. Soc. A* **382**, 2266, p. 20230092.

Bass, S. D. (2024). Spinning protons and gluons in the η', *Int. J. Mod. Phys. A* **39**, 09n10, p. 2441008.

Bass, S. D., De Roeck, A., and Kado, M. (2021). The Higgs boson implications and prospects for future discoveries, *Nature Rev. Phys.* **3**, 9, pp. 608–624.

Bass, S. D. and Doser, M. (2024). Quantum sensing for particle physics, *Nature Rev. Phys.* **6**, 5, pp. 329–339.

Bass, S. D., Harz, J., and Heisenberg, L. (2023). The particle-gravity frontier, *Phil. Trans. Roy. Soc. A* **382**, 2266, p. 20230093.

Bass, S. D. *et al.* (1991). On the infrared contribution to the photon — gluon scattering and the proton spin content, *J. Moscow. Phys. Soc.* **1**, pp. 317–333.

Bass, S. D. and Krzysiak, J. (2020a). The cosmological constant and Higgs mass with emergent gauge symmetries, *Acta Phys. Polon. B* **51**, pp. 1251–1266.

Bass, S. D. and Krzysiak, J. (2020b). Vacuum energy with mass generation and Higgs bosons, *Phys. Lett. B* **803**, p. 135351.

Bass, S. D. and Moskal, P. (2019). η' and η mesons with connection to anomalous glue, *Rev. Mod. Phys.* **91**, 1, p. 015003.

Bass, S. D. and Thomas, A. W. (1993). The EMC spin effect, *J. Phys. G* **19**, pp. 925–956.

Bass, S. D. and Thomas, A. W. (2010). The nucleon's octet axial-charge $g_A^{(8)}$ with chiral corrections, *Phys. Lett. B* **684**, pp. 216–220.

Baudis, L. (2018). The search for Dark Matter, *European Review* **26**, 1, pp. 70–81.

Baudis, L. (2023). Dual-phase xenon time projection chambers for rare-event searches, *Phil. Trans. Roy. Soc. A* **382**, 2266, p. 20230083.

Baumann, D. and Peiris, H. V. (2009). Cosmological inflation: Theory and observations, *Adv. Sci. Lett.* **2**, pp. 105–120.

Bednyakov, A. V. *et al.* (2015). Stability of the electroweak vacuum: Gauge independence and advanced precision, *Phys. Rev. Lett.* **115**, 20, p. 201802.

Bednyakov, A. V., Pikelner, A. F., and Velizhanin, V. N. (2013a). Higgs self-coupling beta-function in the Standard Model at three loops, *Nucl. Phys. B* **875**, pp. 552–565.

Bednyakov, A. V., Pikelner, A. F., and Velizhanin, V. N. (2013b). Yukawa coupling beta-functions in the Standard Model at three loops, *Phys. Lett. B* **722**, pp. 336–340.

Bekenstein, J. D. (1973). Black holes and entropy, *Phys. Rev. D* **7**, pp. 2333–2346.

Belavin, A. A. *et al.* (1975). Pseudoparticle solutions of the Yang-Mills equations, *Phys. Lett. B* **59**, pp. 85–87.

Bell, J. S. (1973). High-energy behaviour of tree diagrams in gauge theories, *Nucl. Phys. B* **60**, pp. 427–436.

Bell, J. S. and Jackiw, R. (1969). A PCAC puzzle: $\pi^0 \to \gamma\gamma$ in the σ model, *Nuovo Cim. A* **60**, pp. 47–61.

Berezhiani, L., Dvali, G., and Sakhelashvili, O. (2022). de Sitter space as a BRST invariant coherent state of gravitons, *Phys. Rev. D* **105**, 2, p. 025022.

Berges, J., Tetradis, N., and Wetterich, C. (2002). Nonperturbative renormalization flow in quantum field theory and statistical physics, *Phys. Rept.* **363**, pp. 223–386.

Bernstein, A. M. and Holstein, B. R. (2013). Neutral pion lifetime measurements and the QCD chiral anomaly, *Rev. Mod. Phys.* **85**, pp. 49–77.

Bertone, G. and Hooper, D. (2018). History of dark matter, *Rev. Mod. Phys.* **90**, 4, p. 045002.

Bertone, G. and Tait, T., M. P. (2018). A new era in the search for dark matter, *Nature* **562**, 7725, pp. 51–56.

Bertone, G. *et al.* (2020). Gravitational wave probes of dark matter: Challenges and opportunities, *SciPost Phys. Core* **3**, p. 007.

Bevan, T. D. C. *et al.* (1997). Momentum creation by vortices in ^{3}he experiments as a model of primordial baryogenesis, *Nature* **386**, pp. 689–692.

Bezrukov, F. *et al.* (2012). Higgs boson mass and new physics, *JHEP* **10**, p. 140.

Bezrukov, F. L. and Shaposhnikov, M. (2008). The Standard Model Higgs boson as the inflaton, *Phys. Lett. B* **659**, pp. 703–706.

Bian, J. *et al.* (2022). Hyper-Kamiokande experiment: A snowmass white paper, in *Snowmass 2021*, arXiv:2203.02029 [hep-ex].

Bird, S. *et al.* (2016). Did LIGO detect dark matter? *Phys. Rev. Lett.* **116**, 20, p. 201301.

Bjorken, J. (2001a). Emergent gauge bosons, in *4th Workshop on What Comes Beyond the Standard Model?*, arXiv:hep-th/0111196.

Bjorken, J. D. (1963). A Dynamical origin for the electromagnetic field, *Annals Phys.* **24**, pp. 174–187.

Bjorken, J. D. (2001b). Standard model parameters and the cosmological constant, *Phys. Rev. D* **64**, p. 085008.

Bjorken, J. D. (2010). Emergent photons and gravitons: The problem of vacuum structure, in *5th Meeting on CPT and Lorentz Symmetry*, arXiv:1008.0033 [hep-ph].

Bjorken, J. D. and Drell, S. D. (1965). *Relativistic quantum fields* (McGraw Hill).

Boccaletti, A. *et al.* (2024). High precision calculation of the hadronic vacuum polarisation contribution to the muon anomaly, arXiv:2407.10913 [hep-lat].

Boehm, C. *et al.* (2021). Eliminating the LIGO bounds on primordial black hole dark matter, *JCAP* **03**, p. 078.

Bond, A. D. and Litim, D. F. (2019). Price of asymptotic safety, *Phys. Rev. Lett.* **122**, 21, p. 211601.

Borsanyi, S. *et al.* (2021). Leading hadronic contribution to the muon magnetic moment from lattice QCD, *Nature* **593**, 7857, pp. 51–55.

Bouchiat, C., Iliopoulos, J., and Meyer, P. (1972). An anomaly free version of Weinberg's model, *Phys. Lett. B* **38**, pp. 519–523.

Branchina, V. and Messina, E. (2013). Stability, Higgs boson mass and new physics, *Phys. Rev. Lett.* **111**, p. 241801.

Brans, C. and Dicke, R. H. (1961). Mach's principle and a relativistic theory of gravitation, *Phys. Rev.* **124**, pp. 925–935.

Brookfield, A. W. *et al.* (2006). Cosmology of mass-varying neutrinos driven by quintessence: theory and observations, *Phys. Rev. D* **73**, p. 083515, [Erratum: *Phys. Rev. D* **76**, 049901 (2007)].

Brown, L. S. (1994). *Quantum Field Theory* (Cambridge University Press).

Buttazzo, D. *et al.* (2013). Investigating the near-criticality of the Higgs boson, *JHEP* **12**, p. 089.

Cabibbo, N. (1963). Unitary symmetry and leptonic decays, *Phys. Rev. Lett.* **10**, pp. 531–533.

Callan, C. G., Jr., Dashen, R. F., and Gross, D. J. (1976). The structure of the gauge theory vacuum, *Phys. Lett. B* **63**, pp. 334–340.

Calmet, X. and Keller, M. (2015). Cosmological evolution of fundamental constants: From theory to experiment, *Mod. Phys. Lett. A* **30**, 22, p. 1540028.

Capozziello, S. and De Laurentis, M. (2011). Extended theories of gravity, *Phys. Rept.* **509**, pp. 167–321.

Caprini, C. (2024). Strong evidence for the discovery of a gravitational wave background, *Nature Rev. Phys.* **6**, 5, pp. 291–293.

Caprini, C. *et al.* (2016). Science with the space-based interferometer eLISA. II: Gravitational waves from cosmological phase transitions, *JCAP* **1604**, p. 001.

Carena, M. *et al.* (2024). Status of Higgs boson physics, in S. Navas *et al.* (eds.), *The Review of Particle Physics (2023)*, [Particle Data Group] (Chapter 11), Phys. Rev. D **110** (2024), 030001.

Carlitz, R. D., Collins, J. C., and Mueller, A. H. (1988). The role of the axial anomaly in measuring spin dependent parton distributions, *Phys. Lett. B* **214**, pp. 229–236.

Carr, B. and Kuhnel, F. (2022). Primordial black holes as dark matter candidates, *SciPost Phys. Lect. Notes* **48**, p. 1.

Carr, B. J. and Rees, M. J. (1979). The anthropic principle and the structure of the physical world, *Nature* **278**, pp. 605–612.

Chacko, Z., Goh, H.-S., and Harnik, R. (2006). The Twin Higgs: Natural electroweak breaking from mirror symmetry, *Phys. Rev. Lett.* **96**, p. 231802.

Chadha, S. and Nielsen, H. B. (1983). Lorentz invariance as a low-energy phenomenon, *Nucl. Phys. B* **217**, pp. 125–144.

Challinor, A. and Peiris, H. (2009). Lecture notes on the physics of cosmic microwave background anisotropies, *AIP Conf. Proc.* **1132**, 1, pp. 86–140.

Chandrasekhar, S. (1985). *The Mathematical Theory of Black Holes* (Oxford University Press).

Chanowitz, M. S. (2005). The No-Higgs signal: Strong WW scattering at the LHC, *Czech. J. Phys.* **55**, pp. B45–B58.

Charlton, M., Eriksson, S., and Shore, G. M. (2020). *Antihydrogen and Fundamental Physics*, Springer Briefs in Physics (Springer).

Chatrchyan, S. *et al.* (2012). Observation of a new boson at a mass of 125 GeV with the CMS experiment at the LHC, *Phys. Lett. B* **716**, pp. 30–61.

Chen, X., Gu, Z. C., and Wen, X. G. (2010). Local unitary transformation, long-range quantum entanglement, wave function renormalization, and topological order, *Phys. Rev. B* **82**, p. 155138.

Chetyrkin, K. G. and Zoller, M. F. (2012). Three-loop β-functions for top-Yukawa and the Higgs self-interaction in the Standard Model, *JHEP* **06**, p. 033.

Chetyrkin, K. G. and Zoller, M. F. (2013). β-function for the Higgs self-interaction in the Standard Model at three-loop level, *JHEP* **04**, p. 091, [Erratum: *JHEP* **09**, 155 (2013)].

Chiba, T. (2011). The constancy of the constants of nature: Updates, *Prog. Theor. Phys.* **126**, pp. 993–1019.

Chkareuli, J. L., Froggatt, C. D., and Nielsen, H. B. (2001). Lorentz invariance and origin of symmetries, *Phys. Rev. Lett.* **87**, p. 091601.

Cichy, K. *et al.* (2015). Non-perturbative test of the Witten-Veneziano formula from lattice QCD, *JHEP* **09**, p. 020.

Close, F. E. (1979). *An Introduction to Quarks and Partons* (Academic N.Y.).

Collins, J. C., Duncan, A., and Joglekar, S. D. (1977). Trace and dilatation anomalies in gauge theories, *Phys. Rev. D* **16**, pp. 438–449.

Collins, J. C., Soper, D. E., and Sterman, G. F. (1989). Factorization of hard processes in QCD, *Adv. Ser. Direct. High Energy Phys.* **5**, pp. 1–91.

Colpi, M. *et al.* (2024). LISA Definition Study Report, arXiv:2402.07571 [astro-ph.CO].

Comparat, D. *et al.* (2023). Experimental perspectives on the matter–antimatter asymmetry puzzle: Developments in electron EDM and H̄ experiments, *Phil. Trans. Roy. Soc. A* **382**, 2266, p. 20230089.

Copeland, E. J., Sami, M., and Tsujikawa, S. (2006). Dynamics of dark energy, *Int. J. Mod. Phys. D* **15**, pp. 1753–1936.

Cornwall, J. M., Levin, D. N., and Tiktopoulos, G. (1973). Uniqueness of spontaneously broken gauge theories, *Phys. Rev. Lett.* **30**, pp. 1268–1270, [Erratum: *Phys. Rev. Lett.* **31**, 572 (1973)].

Cornwall, J. M., Levin, D. N., and Tiktopoulos, G. (1974). Derivation of gauge invariance from high-energy unitarity bounds on the S matrix, *Phys. Rev. D* **10**, pp. 1145–1167, [Erratum: *Phys. Rev. D* **11**, 972 (1975)].

Creutz, M. (1985). *Quarks, Gluons and Lattices* (Cambridge University Press).

Crewther, R. J. (1978). Effects of topological charge in gauge theories, *Acta Phys. Austriaca Suppl.* **19**, pp. 47–153.

Crewther, R. J. (1980). Chiral properties of quantum chromodynamics, *NATO Sci. Ser. B* **55**, pp. 529–590.

Crewther, R. J. *et al.* (1979). Chiral estimate of the electric dipole moment of the neutron in quantum chromodynamics, *Phys. Lett. B* **88**, pp. 123–127, [Erratum: *Phys. Lett. B* **91**, 487 (1980)].

Cronstrom, C. and Mickelsson, J. (1983). On topological boundary characteristics in nonabelian gauge theory, *J. Math. Phys.* **24**, pp. 2528–5231, [Erratum: *J. Math. Phys.* **27**, 419 (1986)].

Csaki, C., Grojean, C., and Terning, J. (2016). Alternatives to an elementary Higgs, *Rev. Mod. Phys.* **88**, 4, p. 045001.

Davies, J. *et al.* (2020). Gauge coupling β functions to four-loop order in the Standard Model, *Phys. Rev. Lett.* **124**, 7, p. 071803.

Degrassi, G. *et al.* (2012). Higgs mass and vacuum stability in the Standard Model at NNLO, *JHEP* **08**, p. 098.

Di Vecchia, P. and Veneziano, G. (1980). Chiral dynamics in the large n limit, *Nucl. Phys. B* **171**, pp. 253–272.

Dirac, P. A. M. (1996). *General Theory of Relativity* (Princeton University Press).

Djukanovic, D., Gegelia, J., and Meißner, U.-G. (2018). Triviality of quantum electrodynamics revisited, *Commun. Theor. Phys.* **69**, 3, pp. 263–265.

Döbrich, B. (2022). Dark matter and axion searches, *PoS* **EPS-HEP2021**, p. 021.

Dokshitzer, Y. L. and Webber, B. R. (1995). Calculation of power corrections to hadronic event shapes, *Phys. Lett. B* **352**, pp. 451–455.

Dolinski, M. J., Poon, A. W. P., and Rodejohann, W. (2019). Neutrinoless double-beta decay: Status and prospects, *Ann. Rev. Nucl. Part. Sci.* **69**, pp. 219–251.

Domcke, V. (2024). Discovery Opportunities with Gravitational Waves — TASI 2024 Lecture Notes, arXiv:2409.08956 [astro-ph.CO].

Dreitlein, J. (1974). Broken symmetry and the cosmological constant, *Phys. Rev. Lett.* **33**, pp. 1243–1244.

Dunkley, J. (2015). Perspective on the cosmic microwave background, *EPL* **111**, 4, p. 49001.

Durrer, R. (2020). *The Cosmic Microwave Background* (Cambridge University Press).

Dvali, G. (2020). *S*-Matrix and anomaly of de Sitter, *Symmetry* **13**, 1, p. 3.

Dvali, G., Gomez, C., and Zell, S. (2017). Quantum break-time of de Sitter, *JCAP* **06**, p. 028.

Dvali, G. R. (1995). Removing the cosmological bound on the axion scale, arXiv:hep-ph/9505253.

Efremov, A. V. and Teryaev, O. V. (1988). Spin structure of the nucleon and triangle anomaly, Dubna preprint JINR-E2-88-287.

Efstathiou, G. (2024). Challenges to the Lambda CDM Cosmology, arXiv:2406.12106 [astro-ph.CO].

Efstathiou, G., Sutherland, W. J., and Maddox, S. J. (1990). The cosmological constant and cold dark matter, *Nature* **348**, pp. 705–707.

Einhorn, M. B. and Jones, D. R. T. (1992). The effective potential and quadratic divergences, *Phys. Rev. D* **46**, pp. 5206–5208.

Einstein, A. (1915). Die Feldgleichungen der Gravitation, *Sitzungsber. Preuss. Akad. Wiss. Berlin (Math. Phys.)* **1915**, pp. 844–847.

Einstein, A. (1917). Kosmologische betrachtungen zur allgemeinen relativitätstheorie, *Sitzungsber. Preuss. Akad. Wiss. Berlin (Math. Phys.)* **1917**, pp. 142–152.

Einstein, A. (1918). Über Gravitationswellen, *Sitzungsber. Preuss. Akad. Wiss. Berlin (Math. Phys.)* **1918**, pp. 154–167.

Einstein, A. (1919). Spielen Gravitationsfelder im aufbau der materiellen Elementarteilchen eine wesentliche rolle? *Sitzungsber. Preuss. Akad. Wiss. Berlin (Math. Phys.)* **1919**, pp. 349–356.

Einstein, A. (1931). Zum kosmologischen problem der allgemeinen relativitaetstheorie, *Sitzungsber. Preuss. Akad. Wiss. Berlin (Math. Phys.)* **1931**, pp. 235–237.

Einstein, A. (1956). *The Meaning of Relativity* (Princeton Univ. Press, Princeton).

Elitzur, S. (1975). Impossibility of spontaneously breaking local symmetries, *Phys. Rev. D* **12**, pp. 3978–3982.

Ellis, J. (2014). The discovery of the gluon, *Int. J. Mod. Phys. A* **29**, 31, p. 1430072.

Ellis, J. *et al.* (2021). Top, Higgs, Diboson and electroweak fit to the Standard Model effective field theory, *JHEP* **04**, p. 279.

Ellis, J. and Wands, D. (2024). Inflation, in S. Navas *et al.* (eds.), *The Review of Particle Physics (2023),* [Particle Data Group] (Chapter 23), Phys. Rev. D **110** (2024), 030001.

Englert, F. and Brout, R. (1964). Broken symmetry and the mass of gauge vector mesons, *Phys. Rev. Lett.* **13**, pp. 321–323.

Escamilla, L. A. *et al.* (2024). The state of the dark energy equation of state circa 2023, *JCAP* **05**, p. 091.

Evans, M. *et al.* (2021). A horizon study for cosmic explorer: Science, observatories, and community, arXiv:2109.09882 [astro-ph.IM].

Fan, X. *et al.* (2023). Measurement of the electron magnetic moment, *Phys. Rev. Lett.* **130**, 7, p. 071801.

Fardon, R., Nelson, A. E., and Weiner, N. (2004). Dark energy from mass varying neutrinos, *JCAP* **10**, p. 005.

Ferrarese, L. (2002). Beyond the bulge: A fundamental relation between supermassive black holes and dark matter halos, *Astrophys. J.* **578**, pp. 90–97.

Ferrarese, L. and Ford, H. (2005). Supermassive black holes in galactic nuclei: Past, present and future research, *Space Sci. Rev.* **116**, pp. 523–624.

Ferrarese, L. *et al.* (2006). A fundamental relation between compact stellar nuclei, supermassive black holes, and their host galaxies, *Astrophys. J. Lett.* **644**, pp. L21–L24.

Feynman, R. P. (1963). Quantum theory of gravitation, *Acta Phys. Polon.* **24**, pp. 697–722.

Feynman, R. P. (1999). The reason for anti-particles, in *Elementary Particles and the Laws of Physics The 1986 Dirac Memorial Lectures* (Cambridge University Press).

Feynman, R. P., Morinigo, F. B., and Wagner, W. G. (1995). *Feynman Lectures on Gravitation* (Addison-Wesley).

Ford, C. *et al.* (1993). The effective potential and the renormalization group, *Nucl. Phys. B* **395**, pp. 17–34.

Forster, D., Nielsen, H. B., and Ninomiya, M. (1980). Dynamical stability of local gauge symmetry: Creation of light from chaos, *Phys. Lett. B* **94**, pp. 135–140.

Freedman, W. L. (2017). Cosmology at a crossroads, *Nature Astron.* **1**, p. 0121.

Freedman, W. L. *et al.* (2024). Status report on the chicago-carnegie hubble program (CCHP): Three independent astrophysical determinations of the hubble constant using the James Webb space telescope, arXiv:2408.06153 [astro-ph.CO].

Frésard, R. (2015). The slave-boson approach to correlated fermions, in E. Pavarini, E. Koch, and P. Coleman (eds.), *Many-Body Physics: From Kondo to Hubbard Modeling and Simulation*, chap. 9 (Verlag des Forschungszentrum Jülich, Jülich), pp. 1–36.

Frieman, J., Turner, M., and Huterer, D. (2008). Dark energy and the accelerating universe, *Ann. Rev. Astron. Astrophys.* **46**, pp. 385–432.

Fritzsch, H. (2010). The fundamental constants in physics and their possible time variation, *Nucl. Phys. B Proc. Suppl.* **203-204**, pp. 3–17.

Fritzsch, H. (2012). The size of the weak bosons, *Phys. Lett. B* **712**, pp. 231–232.

Fritzsch, H. and Gell-Mann, M. (1972). Current algebra: Quarks and what else? *eConf* **C720906V2**, pp. 135–165.

Fritzsch, H., Gell-Mann, M., and Leutwyler, H. (1973). Advantages of the color octet gluon picture, *Phys. Lett. B* **47**, pp. 365–368.

Fritzsch, H. and Mandelbaum, G. (1981). Weak interactions as manifestations of the substructure of leptons and quarks, *Phys. Lett. B* **102**, pp. 319–322.

Fritzsch, H. and Xing, Z.-z. (1998). Large leptonic flavor mixing and the mass spectrum of leptons, *Phys. Lett. B* **440**, pp. 313–318.

Froggatt, C. D. and Nielsen, H. B. (1979). Hierarchy of quark masses, Cabibbo angles and CP violation, *Nucl. Phys. B* **147**, pp. 277–298.

Frohlich, J., Morchio, G., and Strocchi, F. (1980). Higgs phenomenon without a symmetry breaking order parameter, *Phys. Lett. B* **97**, pp. 249–252.

Frohlich, J., Morchio, G., and Strocchi, F. (1981). Higgs phenomenon without symmetry breaking order parameter, *Nucl. Phys. B* **190**, pp. 553–582.

Gasser, J. and Leutwyler, H. (1982). Quark masses, *Phys. Rept.* **87**, pp. 77–169.

Gehrmann, T. and Malaescu, B. (2022). Precision QCD physics at the LHC, *Ann. Rev. Nucl. Part. Sci.* **72**, pp. 233–258.

Gell-Mann, M., Oakes, R. J., and Renner, B. (1968). Behavior of current divergences under SU(3) x SU(3), *Phys. Rev.* **175**, pp. 2195–2199.

Gell-Mann, M., Ramond, P., and Slansky, R. (1979). Complex spinors and unified theories, *Conf. Proc. C* **790927**, pp. 315–321.

Genzel, R., Eisenhauer, F., and Gillessen, S. (2024). Experimental studies of black holes: Status and future prospects, *Astron. Astrophys. Rev.* **32**, 1, p. 3.

Genzel, R. *et al.* (2017). Strongly baryon-dominated disk galaxies at the peak of galaxy formation ten billion years ago, *Nature* **543**, pp. 397–401.

Glashow, S. L. (1961). Partial symmetries of weak interactions, *Nucl. Phys.* **22**, pp. 579–588.

Glashow, S. L. (1980). The future of elementary particle physics, *NATO Sci. Ser. B* **61**, pp. 687–713.

Gockeler, M. *et al.* (1998). Is there a Landau pole problem in QED? *Phys. Rev. Lett.* **80**, pp. 4119–4122.

Goldstone, J. (1961). Field theories with superconductor solutions, *Nuovo Cim.* **19**, pp. 154–164.

Goldstone, J., Salam, A., and Weinberg, S. (1962). Broken symmetries, *Phys. Rev.* **127**, pp. 965–970.

Gomes, P. R. S. (2016). Aspects of emergent symmetries, *Int. J. Mod. Phys. A* **31**, 10, p. 1630009.

Gómez, C. and Letschka, R. (2020). Masses and electric charges: Gauge anomalies and anomalous thresholds, *Eur. Phys. J. C* **80**, 10, p. 946.

Green, A. M. (2024). Primordial black holes as a dark matter candidate — a brief overview, *Nucl. Phys. B* **1003**, p. 116494.

Green, A. M. and Kavanagh, B. J. (2021). Primordial black holes as a dark matter candidate, *J. Phys. G* **48**, 4, p. 043001.

Gribov, V. N. (1977). Instability of nonabelian gauge theories and impossibility of choice of coulomb gauge, SLAC-TRANS-0176 (1977), in J. Nyiri (ed.), *The Gribov Theory of Quark Confinement* (World Scientific, 2001), pp. 24–38.

Gribov, V. N. (1978). Quantization of nonabelian gauge theories, *Nucl. Phys. B* **139**, pp. 1–19.

Gribov, V. N. (1981). Anomalies, as a manifestation of the high momentum collective motion in the vacuum. Budapest preprint KFKI-1981-66 (1981), in J. Nyiri (ed.), *The Gribov Theory of Quark Confinement* (World Scientific, 2001), pp. 74–91.

Gribov, V. N. (1982). Local confinement of charge in massless QED, *Nucl. Phys. B* **206**, pp. 103–131.

Gribov, V. N. (1987). A new hypothesis on the nature of quark and gluon confinement, *Phys. Scripta T* **15**, pp. 164–168.

Gribov, V. N. (1999). The theory of quark confinement, *Eur. Phys. J. C* **10**, pp. 91–105.

Grimm, N., Bonvin, C., and Tutusaus, I. (2024). New measurements of E_G: Testing General Relativity with the Weyl potential and galaxy velocities, arXiv:2403.13709 [astro-ph.CO].

Gross, D. J. and Jackiw, R. (1972). Effect of anomalies on quasirenormalizable theories, *Phys. Rev. D* **6**, pp. 477–493.

Gross, D. J. and Wilczek, F. (1973). Ultraviolet behavior of nonabelian gauge theories, *Phys. Rev. Lett.* **30**, pp. 1343–1346.

Grosse, H., Langmann, E., and Raschhofer, E. (1997). The Luttinger-Schwinger model, *Annals Phys.* **253**, pp. 310–331.

Grzadkowski, B. *et al.* (2010). Dimension-six terms in the Standard Model Lagrangian, *JHEP* **10**, p. 085.

Guralnik, G. S., Hagen, C. R., and Kibble, T. W. B. (1964). Global conservation laws and massless particles, *Phys. Rev. Lett.* **13**, pp. 585–587.

Halimeh, J. C. and Hauke, P. (2022). Stabilizing gauge theories in quantum simulators: A brief review, arXiv:2204.13709 [cond-mat.quant-gas].

Hamada, Y., Kawai, H., and Oda, K.-y. (2013). Bare Higgs mass at Planck scale, *Phys. Rev. D* **87**, 5, p. 053009, [Erratum: *Phys. Rev. D* **89**, 059901 (2014)].

Hambye, T. and Riesselmann, K. (1997). Matching conditions and Higgs mass upper bounds revisited, *Phys. Rev. D* **55**, pp. 7255–7262.

Han, T. H. *et al.* (2012). Fractionalized excitations in the spin-liquid state of a kagome-lattice antiferromagnet, *Nature* **492**, p. 406.

Hart, L. and Chluba, J. (2018). New constraints on time-dependent variations of fundamental constants using Planck data, *Mon. Not. Roy. Astron. Soc.* **474**, 2, pp. 1850–1861.

Hawking, S. W. (1974). Black hole explosions, *Nature* **248**, pp. 30–31.

Hawking, S. W. (1975). Particle creation by black holes, *Commun. Math. Phys.* **43**, pp. 199–220, [Erratum: *Commun. Math. Phys.* **46**, 206 (1976)].

Hayrapetyan, A. *et al.* (2024a). Review of top quark mass measurements in CMS, arXiv:2403.01313 [hep-ex].

Hayrapetyan, A. *et al.* (2024b). Stairway to discovery: A report on the CMS programme of cross section measurements from millibarns to femtobarns, arXiv:2405.18661 [hep-ex].

Higgs, P. W. (1964a). Broken symmetries and the masses of gauge bosons, *Phys. Rev. Lett.* **13**, pp. 508–509.

Higgs, P. W. (1964b). Broken symmetries, massless particles and gauge fields, *Phys. Lett.* **12**, pp. 132–133.

Higgs, P. W. (1966). Spontaneous symmetry breakdown without massless bosons, *Phys. Rev.* **145**, pp. 1156–1163.

Hiller, G. *et al.* (2024). Vacuum Stability in the Standard Model and Beyond, arXiv:2401.08811 [hep-ph].

Hoang, A. H. (2020). What is the top quark mass? *Ann. Rev. Nucl. Part. Sci.* **70**, pp. 225–255.

Hook, A. (2019). TASI lectures on the strong CP problem and axions, *PoS* **TASI2018**, p. 004.

Huston, J. H., Rabbertz, K., and Zanderighi, G. (2024). Quantum chromodynamics, in S. Navas *et al.* (eds.), *The Review of Particle Physics (2023)*, [Particle Data Group] (Chapter 9), Phys. Rev. D **110** (2024), 030001.

Ioffe, B. L. (2006). Axial anomaly: The modern status, *Int. J. Mod. Phys. A* **21**, pp. 6249–6266.

Isidori, G., Wilsch, F., and Wyler, D. (2024). The standard model effective field theory at work, *Rev. Mod. Phys.* **96**, 1, p. 015006.

Itzykson, C. and Zuber, J. B. (1980). *Quantum Field Theory*, International Series In Pure and Applied Physics (McGraw-Hill, New York).

Jackiw, R. and Johnson, K. (1969). Anomalies of the axial vector current, *Phys. Rev.* **182**, pp. 1459–1469.

Jackiw, R. and Rebbi, C. (1976). Vacuum periodicity in a Yang-Mills quantum theory, *Phys. Rev. Lett.* **37**, pp. 172–175.

Jaffe, R. L. (1996). Spin, twist and hadron structure in deep inelastic processes, in *Ettore Majorana International School of Nucleon Structure: 1st Course: The Spin Structure of the Nucleon*, pp. 42–129, arXiv:hep-ph/9602236.

Jaffe, R. L. (2005). The Casimir effect and the quantum vacuum, *Phys. Rev. D* **72**, p. 021301.

Jaffe, R. L. and Manohar, A. (1990). The g_1 problem: Deep inelastic electron scattering and the spin of the proton, *Nucl. Phys. B* **337**, pp. 509–546.

Jakobs, K. and Zanderighi, G. (2023). The profile of the Higgs boson: Status and prospects, *Phil. Trans. Roy. Soc. A* **382**, 2266, p. 20230087.

Janot, P. and Jadach, S. (2020). Improved Bhabha cross section at LEP and the number of light neutrino species, *Phys. Lett. B* **803**, p. 135319.

Jarlskog, C. (1985). A basis independent formulation of the connection between quark mass matrices, CP violation and experiment, *Z. Phys. C* **29**, pp. 491–497.

Jegerlehner, F. (1978). The vector boson and graviton propagators in the presence of multipole forces, *Helv. Phys. Acta* **51**, pp. 783–792.

Jegerlehner, F. (1990). Renormalizing the standard model, *Conf. Proc. C* **900603**, pp. 476–590.

Jegerlehner, F. (1998). The 'Ether world' and elementary particles, in D. L. H. Dorn and G. Weigt (eds.), *31st International Ahrenshoop Symposium on the Theory of Elementary Particles* (Wiley-VCH), pp. 386–392, arXiv:hep-th/9803021.

Jegerlehner, F. (2014a). About the role of the Higgs boson in the evolution of the early universe, *Acta Phys. Polon. B* **45**, 7, pp. 1393–1413.

Jegerlehner, F. (2014b). Higgs inflation and the cosmological constant, *Acta Phys. Polon. B* **45**, 6, pp. 1215–1254.

Jegerlehner, F. (2014c). The Standard model as a low-energy effective theory: What is triggering the Higgs mechanism? *Acta Phys. Polon. B* **45**, 6, pp. 1167–1214.

Jegerlehner, F. (2017). *The Anomalous Magnetic Moment of the Muon*, Vol. 274 (Springer, Cham).

Jegerlehner, F. (2019). The hierarchy problem and the cosmological constant problem revisited — A new view on the SM of particle physics, *Found. Phys.* **49**, 9, pp. 915–971.

Jegerlehner, F. (2021). The standard model of particle physics as a conspiracy theory and the possible role of the Higgs boson in the evolution of the early universe, *Acta Phys. Polon. B* **52**, 6–7, pp. 575–605.

Jegerlehner, F. (2023). Is the Higgs Boson the Master of the Universe? arXiv:2305.01326 [hep-ph].

Jegerlehner, F. and Fleischer, J. (1985). High-energy behavior of the electromagnetic singlet current in the Glashow-Weinberg-Salam model, *Phys. Lett. B* **151**, pp. 65–68.

Jegerlehner, F. and Fleischer, J. (1986). Singlet form-factors and local observables in the Glashow-Weinberg-Salam model, *Acta Phys. Polon. B* **17**, pp. 709–733.

Jegerlehner, F., Kalmykov, M. Y., and Kniehl, B. A. (2013). On the difference between the pole and the $\overline{MS}$ masses of the top quark at the electroweak scale, *Phys. Lett. B* **722**, pp. 123–129.

Jones, D. R. T. (1982). The two loop beta function for a $G(1) \times G(2)$ gauge theory, *Phys. Rev. D* **25**, pp. 581–582.

Jones, D. R. T. (2013). Comment on "Bare Higgs mass at Planck scale", *Phys. Rev. D* **88**, 9, p. 098301.

Kamenshchik, A. Y. *et al.* (2018). Pauli–Zeldovich cancellation of the vacuum energy divergences, auxiliary fields and supersymmetry, *Eur. Phys. J. C* **78**, 3, p. 200.

Kamionkowski, M. and Kovetz, E. D. (2016). The quest for B modes from inflationary gravitational waves, *Ann. Rev. Astron. Astrophys.* **54**, pp. 227–269.

Kaplan, D. B. (1991). Flavor at SSC energies: A new mechanism for dynamically generated fermion masses, *Nucl. Phys. B* **365**, pp. 259–278.

Kawasaki, M. and Nakayama, K. (2013). Axions: theory and cosmological role, *Ann. Rev. Nucl. Part. Sci.* **63**, pp. 69–95.

Kharzeev, D. E. (2011). Axial anomaly, Dirac sea, and the chiral magnetic effect, in Y. L. Dokshitzer, P. Levai, and J. Nyiri (eds.), *Gribov-80 Memorial Volume* (World Scientific), pp. 293–306.

Kibble, T. W. B. (1961). Lorentz invariance and the gravitational field, *J. Math. Phys.* **2**, pp. 212–221.

Kibble, T. W. B. (1967). Symmetry breaking in non-Abelian gauge theories, *Phys. Rev.* **155**, pp. 1554–1561.

Kibble, T. W. B. (2014). Spontaneous symmetry breaking in gauge theories, *Phil. Trans. Roy. Soc. A* **373**, 2032, p. 20140033.

Kitaev, A. and Preskill, J. (2006). Topological entanglement entropy, *Phys. Rev. Lett.* **96**, p. 110404.

Klevansky, S. P. (1992). The Nambu-Jona-Lasinio model of quantum chromodynamics, *Rev. Mod. Phys.* **64**, pp. 649–708.

Klinkhamer, F. R. and Volovik, G. E. (2017). Dark matter from dark energy in q-theory, *JETP Lett.* **105**, 2, pp. 74–77.

Kniehl, B. A., Pikelner, A. F., and Veretin, O. L. (2015). Two-loop electroweak threshold corrections in the Standard Model, *Nucl. Phys. B* **896**, pp. 19–51.

Kniehl, B. A., Pikelner, A. F., and Veretin, O. L. (2016). mr: a C++ library for the matching and running of the Standard Model parameters, *Comput. Phys. Commun.* **206**, pp. 84–96.

Kobayashi, M. and Maskawa, T. (1973). CP violation in the renormalizable theory of weak interaction, *Prog. Theor. Phys.* **49**, pp. 652–657.

Kogut, J. B., Dagotto, E., and Kocic, A. (1988). A new phase of quantum electrodynamics: A nonperturbative fixed point in four-dimensions, *Phys. Rev. Lett.* **60**, pp. 772–775.

Kolb, E. W., Chung, D. J. H., and Riotto, A. (1999). WIMPzillas! *AIP Conf. Proc.* **484**, 1, pp. 91–105.

Komatsu, E. (2022). New physics from the polarized light of the cosmic microwave background, *Nature Rev. Phys.* **4**, 7, pp. 452–469.

Kragh, H. S. and Overduin, J. M. (2014). *The Weight of the Vacuum: A Scientific History of Dark Energy*, Springerbriefs in Physics (Springer, Heidelberg).

Kramer, M. *et al.* (2006). Tests of general relativity from timing the double pulsar, *Science* **314**, pp. 97–102.

Kramer, M. *et al.* (2021). Strong-Field gravity tests with the double pulsar, *Phys. Rev. X* **11**, 4, p. 041050.

Kraus, E. (1998). Renormalization of the electroweak Standard Model to all orders, *Annals Phys.* **262**, pp. 155–259.

Križan, P. (2023). Flavour physics as a window to new physics searches, *Phil. Trans. Roy. Soc. A* **382**, 2266, p. 20230088.

Kun, E. *et al.* (2023). Searching for temporary gamma-ray dark blazars associated with IceCube neutrinos, *Astron. Astrophys.* **679**, p. A46.

Lahav, O. and Liddle, A. R. (2024). Cosmological parameters, in S. Navas *et al.* (eds.), *The Review of Particle Physics (2023)*, [Particle Data Group] (Chapter 25), Phys. Rev. D **110** (2024), 030001.

Landau, L. D., Abrikosov, A., and Halatnikov, L. (1956). On the quantum theory of fields, *Nuovo Cim. Suppl.* **3**, pp. 80–104.

Landau, L. D. and Pomeranchuk, I. Y. (1955). On point interactions in quantum electrodynamics, *Dokl. Akad. Nauk SSSR* **102**, 3, pp. 489–492.

Lange, R. *et al.* (2021). Improved limits for violations of local position invariance from atomic clock comparisons, *Phys. Rev. Lett.* **126**, 1, p. 011102.

Lee, A. J. *et al.* (2024). The Chicago-Carnegie Hubble Program: The JWST J-region Asymptotic Giant Branch (JAGB) Extragalactic Distance Scale, arXiv:2408.03474 [astro-ph.GA].

Lee, D. M. (1997). The extraordinary phases of liquid He-3, *Rev. Mod. Phys.* **69**, pp. 645–666, [Erratum: *Rev. Mod. Phys.* **70**, 319–319 (1998)].

Lee, J. G. *et al.* (2020). New test of the gravitational $1/r^2$ law at separations down to 52 μm, *Phys. Rev. Lett.* **124**, 10, p. 101101.

Lesgourgues, J. and Verde, L. (2024). Neutrinos in cosmology, in S. Navas *et al.* (eds.), *The Review of Particle Physics (2024)*, [Particle Data Group] (Chapter 26), Phys. Rev. D **110** (2024), 030001.

Leutwyler, H. (1998). On the 1/N expansion in chiral perturbation theory, *Nucl. Phys. B Proc. Suppl.* **64**, pp. 223–231.

Leutwyler, H. (2013). The mass of the two lightest quarks, *Mod. Phys. Lett. A* **28**, p. 1360014.

Levin, M. and Wen, X.-G. (2006a). Detecting topological order in a ground state wave function, *Phys. Rev. Lett.* **96**, p. 110405.

Levin, M. and Wen, X.-G. (2006b). Quantum ether: Photons and electrons from a rotor model, *Phys. Rev. B* **73**, p. 035122.

Levin, M. A. and Wen, X.-G. (2005a). Colloquium: Photons and electrons as emergent phenomena, *Rev. Mod. Phys.* **77**, pp. 871–879.

Levin, M. A. and Wen, X.-G. (2005b). String net condensation: A Physical mechanism for topological phases, *Phys. Rev. B* **71**, p. 045110.

Livio, M. and Rees, M. J. (2005). Anthropic reasoning, *Science* **309**, pp. 1022–1023.

Llewellyn Smith, C. H. (1973). High-Energy behavior and gauge symmetry, *Phys. Lett. B* **46**, pp. 233–236.

Llewellyn Smith, C. H. (1989). Quark correlation functions and deep inelastic scattering, in B. Castel and P. J. O'Donnell (eds.), *Proc. CAP-NSERC Summer Institute on Particles and Fields* (World Scientific), p. 309.

Loebbert, F. (2008). The Weinberg-Witten theorem on massless particles: An Essay, *Annalen Phys.* **17**, pp. 803–829.

Loll, R., Fabiano, G., Frattulillo, D., and Wagner, F. (2022). Quantum gravity in 30 questions, *PoS* **CORFU2021**, p. 316.

Luscher, M. (2002). Chiral gauge theories revisited, *Subnucl. Ser.* **38**, pp. 41–89.

Luty, M. A., Polchinski, J., and Rattazzi, R. (2013). The a-theorem and the asymptotics of 4D quantum field theory, *JHEP* **01**, p. 152.

Lyth, D. H. (1997). What would we learn by detecting a gravitational wave signal in the cosmic microwave background anisotropy? *Phys. Rev. Lett.* **78**, pp. 1861–1863.

Machacek, M. E. and Vaughn, M. T. (1983). Two loop renormalization group equations in a general quantum field theory. 1. Wave function renormalization, *Nucl. Phys. B* **222**, pp. 83–103.

Machacek, M. E. and Vaughn, M. T. (1984). Two loop renormalization group equations in a general quantum field theory. 2. Yukawa couplings, *Nucl. Phys. B* **236**, pp. 221–232.

Machacek, M. E. and Vaughn, M. T. (1985). Two loop renormalization group equations in a general quantum field theory. 3. Scalar quartic couplings, *Nucl. Phys. B* **249**, pp. 70–92.

Maggiore, M. (2007). *Gravitational Waves. Vol. 1: Theory and Experiments* (Oxford University Press).

Maggiore, M. (2018). *Gravitational Waves. Vol. 2: Astrophysics and Cosmology* (Oxford University Press).

Maggiore, M. *et al.* (2020). Science case for the Einstein telescope, *JCAP* **03**, p. 050.

Maki, Z., Nakagawa, M., and Sakata, S. (1962). Remarks on the unified model of elementary particles, *Prog. Theor. Phys.* **28**, pp. 870–880.

Mandula, J. E. (1977). Color screening by a Yang-Mills instability, *Phys. Lett. B* **67**, pp. 175–178.

Martin, J. (2012). Everything you always wanted to know about the cosmological constant problem (But were afraid to ask), *Comptes Rendus Physique* **13**, pp. 566–665.

Masina, I. (2013). Higgs boson and top quark masses as tests of electroweak vacuum stability, *Phys. Rev. D* **87**, 5, p. 053001.

Masina, I. and Quiros, M. (2013). On the Veltman condition, the hierarchy problem and high-scale supersymmetry, *Phys. Rev. D* **88**, p. 093003.

Mattingly, A. C. and Stevenson, P. M. (1994). Optimization of R(e+ e-) and 'freezing' of the QCD couplant at low-energies, *Phys. Rev. D* **49**, pp. 437–450.

Mayer, A. C. *et al.* (2022). ΛCDM with baryons versus MOND: The time evolution of the universal acceleration scale in the Magneticum simulations, *Mon. Not. Roy. Astron. Soc.* **518**, 1, pp. 257–269.

McGaugh, S. S. *et al.* (2000). The baryonic Tully-Fisher relation, *Astrophys. J. Lett.* **533**, pp. L99–L102.

Melnitchouk, W., Thomas, A. W., and Signal, A. I. (1991). Gottfried sum rule and the shape of $F_2^p - F_2^n$, *Z. Phys. A* **340**, pp. 85–92.

Milgrom, M. (1983). A modification of the newtonian dynamics: Implications for galaxies, *Astrophys. J.* **270**, pp. 371–383.

Milgrom, M. (2020). The a_0 — cosmology connection in MOND, arXiv:2001.09729 [astro-ph.GA].

Minkowski, P. (1977). $\mu \to e\gamma$ at a Rate of one out of 10^9 muon decays? *Phys. Lett. B* **67**, pp. 421–428.

Misner, C. W., Thorne, K. S., and Wheeler, J. A. (1973). *Gravitation* (W. H. Freeman, San Francisco).

Moessner, R. and Moore, J. E. (2021). *Topological Phases of Matter* (Cambridge University Press).

Mohapatra, R. N. and Senjanovic, G. (1980). Neutrino mass and spontaneous parity nonconservation, *Phys. Rev. Lett.* **44**, pp. 912–915.

Morchio, G. and Strocchi, F. (1986). Confinement of massless charged particles in QED in four-dimensions and of charged particles in QED in three-dimensions, *Annals Phys.* **172**, pp. 267–279.

Morel, L., Yao, Z., Cladé, P., and Guellati-Khélifa, S. (2020). Determination of the fine-structure constant with an accuracy of 81 parts per trillion, *Nature* **588**, 7836, pp. 61–65.

Morrissey, D. E. and Ramsey-Musolf, M. J. (2012). Electroweak baryogenesis, *New J. Phys.* **14**, p. 125003.

Muta, T. (1987). *Foundations of Quantum Chromodynamics: An Introduction to Perturbative Methods in Gauge Theories*, *World Scientific Lecture Notes in Physics*, Vol. 5 (World Scientific).

Nakamura, Y. and Schierholz, G. (2023). The strong CP problem solved by itself due to long-distance vacuum effects, *Nucl. Phys. B* **986**, p. 116063.

Nakayama, Y. (2015). Scale invariance vs conformal invariance, *Phys. Rept.* **569**, pp. 1–93.

Nambu, Y. (1960). Quasiparticles and gauge invariance in the theory of superconductivity, *Phys. Rev.* **117**, pp. 648–663.

Nambu, Y. and Jona-Lasinio, G. (1961). Dynamical model of elementary particles based on an analogy with superconductivity. 1. *Phys. Rev.* **122**, pp. 345–358.

Navas, S. *et al.* (2024). Review of particle physics, *Phys. Rev. D* **110**, p. 030001.

Nielsen, H. B. and Ninomiya, M. (1978). Beta function in a noncovariant Yang-Mills theory, *Nucl. Phys. B* **141**, pp. 153–177.

Osheroff, D. D. (1997). Superfluidity in He-3: Discovery and understanding, *Rev. Mod. Phys.* **69**, pp. 667–682.

Padmanabhan, T. (2003). Cosmological constant: The Weight of the vacuum, *Phys. Rept.* **380**, pp. 235–320.

Palacios, P. (2022). *Emergence and Reduction in Physics* (Cambridge University Press).

Parisi, G. and Wu, Y.-s. (1981). Perturbation theory without gauge fixing, *Sci. Sin.* **24**, pp. 483–496.

Parker, R. H. *et al.* (2018). Measurement of the fine-structure constant as a test of the Standard Model, *Science* **360**, pp. 191–195.

Pauli, W. (1933). *Die allgemeinen Prinzipien der Wellenmechanik*, Handbuch der Physik, vol. xxiv (1933). in N. Straumann (ed.), New edition, Springer Berlin Heidelberg (1990); see Appendix III, p. 202., doi:10.1007/978-3-642-61287-9.

Pauli, W. (1971). *Pauli Lectures on Physics: Vol. 6, Selected Topics in Field Quantization* (MIT Press).

Peccei, R. D. (1999). Discrete and global symmetries in particle physics, *Lect. Notes Phys.* **521**, pp. 1–50.

Peccei, R. D. and Quinn, H. R. (1977). CP conservation in the presence of instantons, *Phys. Rev. Lett.* **38**, pp. 1440–1443.

Peebles, P. J. E. and Ratra, B. (1988). Cosmology with a time variable cosmological constant, *Astrophys. J. Lett.* **325**, pp. L17–L20.

Peebles, P. J. E. and Ratra, B. (2003). The cosmological constant and dark energy, *Rev. Mod. Phys.* **75**, pp. 559–606.

Pendlebury, J. M. *et al.* (2015). Revised experimental upper limit on the electric dipole moment of the neutron, *Phys. Rev. D* **92**, 9, p. 092003.

Penrose, R. (1969). Gravitational collapse: The role of general relativity, *Riv. Nuovo Cim.* **1**, pp. 252–276.

Penrose, R. (1989). Difficulties with inflationary cosmology, *Annals of the New York Academy of Sciences* **571**, 1, pp. 249–264.

Penrose, R. (2004). *The Road to Reality: A Complete Guide to the Laws of the Universe* (Jonathan Cape).

Percacci, R. (2009). Asymptotic safety, in D. Oriti (ed.), *Approaches to Quantum Gravity: Towards a New Understanding of Space, Time and Matter* (Cambridge University Press), pp. 111–128.

Perlmutter, S. *et al.* (1999). Measurements of Ω and Λ from 42 high redshift supernovae, *Astrophys. J.* **517**, pp. 565–586.

Peskin, M. E. and Schroeder, D. V. (1995). *An Introduction to Quantum Field Theory* (Addison-Wesley, Reading, USA).

Pikovski, I. *et al.* (2012). Probing Planck-scale physics with quantum optics, *Nature Phys.* **8**, pp. 393–397.

Pokorski, S. (2000). *Gauge Field Theories, 2nd edition* (Cambridge University Press).

Pokorski, S. (2016). Physics beyond the Standard Model in hadronic collisions, *Acta Phys. Polon. B* **47**, pp. 1767–1780.

Pokorski, S. (2023). After the Higgs boson discovery: A turning point in particle physics, *Phil. Trans. Roy. Soc. A* **382**, 2266, p. 20230090.

Polchinski, J. (1988). Scale and conformal invariance in quantum field theory, *Nucl. Phys. B* **303**, pp. 226–236.

Politzer, H. D. (1973). Reliable perturbative results for strong interactions? *Phys. Rev. Lett.* **30**, pp. 1346–1349.

Polyakov, A. M. (1977). Quark confinement and topology of gauge groups, *Nucl. Phys. B* **120**, pp. 429–458.

Polyakov, A. M. (2010). Decay of vacuum energy, *Nucl. Phys. B* **834**, pp. 316–329.

Polyakov, A. M. (2012). Infrared instability of the de Sitter space, arXiv:1209.4135 [hep-th].

Pontecorvo, B. (1957). Inverse beta processes and nonconservation of lepton charge, *Zh. Eksp. Teor. Fiz.* **34**, pp. 247–248.

Powell, B. J. (2020). Emergent particles and gauge fields in quantum matter, *Contemp. Phys.* **61**, 2, pp. 96–131.

Rehn, J. and Moessner, R. (2016). Maxwell electromagnetism as an emergent phenomenon in condensed matter, *Phil. Trans. Roy. Soc. A* **374**, 2075, p. 20160093.

Riess, A. G. (2019). The expansion of the universe is faster than expected, *Nature Rev. Phys.* **2**, 1, pp. 10–12.

Riess, A. G. *et al.* (2022). Cluster cepheids with high precision Gaia parallaxes, low zero-point uncertainties, and Hubble space telescope photometry, *Astrophys. J.* **938**, 1, p. 36.

Riess, A. G. *et al.* (1998). Observational evidence from supernovae for an accelerating universe and a cosmological constant, *Astron. J.* **116**, pp. 1009–1038.

Ringwald, A. (2015). The hunt for axions, *PoS* **NEUTEL2015**, p. 021.

Roberts, R. G. (1990). *The Structure of the Proton: Deep Inelastic Scattering* (Cambridge University Press).

Rosenberg, L. J., Rybka, G., and Safdi, B. (2024). Axions and other similar particles, in S. Navas *et al.* (eds.), *The Review of Particle Physics (2023)*, [Particle Data Group] (Chapter 90), Phys. Rev. D **110** (2024), 030001.

Ross, G. G. (2014). SUSY: Quo Vadis? *Eur. Phys. J. C* **74**, p. 2699.

Ross, G. G. and Roberts, R. G. (1992). Minimal supersymmetric unification predictions, *Nucl. Phys. B* **377**, pp. 571–592.

Roussy, T. S. *et al.* (2023). An improved bound on the electron's electric dipole moment, *Science* **381**, 6653, p. adg4084.

Rubio, J. (2019). Higgs inflation, *Front. Astron. Space Sci.* **5**, p. 50.

Sachdev, S. (2016). Emergent gauge fields and the high temperature superconductors, *Phil. Trans. Roy. Soc. A* **374**, p. 20150248.

Sachdev, S. (2019). Topological order, emergent gauge fields, and Fermi surface reconstruction, *Rept. Prog. Phys.* **82**, 1, p. 014001.

Safronova, M. S. (2019). The search for variation of fundamental constants with clocks, *Annalen Phys.* **531**, 5, p. 1800364.

Sahni, V. and Starobinsky, A. A. (2000). The case for a positive cosmological Lambda term, *Int. J. Mod. Phys. D* **9**, pp. 373–444.

Sakharov, A. D. (1967). Violation of CP invariance, C asymmetry, and baryon asymmetry of the universe, *Pisma Zh. Eksp. Teor. Fiz.* **5**, pp. 32–35.

Salam, A. (1968). Weak and electromagnetic interactions, *Proceedings of the 8th Nobel symposium, Ed. N. Svartholm, Almqvist and Wiskell, 1968,* Conf. Proc. **C680519**, pp. 367–377.

Schael, S. *et al.* (2006). Precision electroweak measurements on the Z resonance, *Phys. Rept.* **427**, pp. 257–454.

Schöneberg, N., Franco Abellán, G., Pérez Sánchez, A., Witte, S. J., Poulin, V., and Lesgourgues, J. (2022). The H_0 Olympics: A fair ranking of proposed models, *Phys. Rept.* **984**, pp. 1–55.

Sciama, D. W. (1964). The physical structure of general relativity, *Rev. Mod. Phys.* **36**, pp. 463–469, [Erratum: *Rev. Mod. Phys.* **36**, 1103–1103 (1964)].

Siegel, D. M. (2022). r-Process nucleosynthesis in gravitational-wave and other explosive astrophysical events, *Nature Rev. Phys.* **4**, 5, pp. 306–318.

Seljak, U. and Zaldarriaga, M. (1997). Signature of gravity waves in polarization of the microwave background, *Phys. Rev. Lett.* **78**, pp. 2054–2057.

Servant, G. (2018). The serendipity of electroweak baryogenesis, *Phil. Trans. Roy. Soc. A* **376**, 2114, p. 20170124.

Shifman, M. (2020). Musings on the current status of HEP, *Mod. Phys. Lett. A* **35**, 07, p. 2030003.

Shifman, M. A. (1991). Anomalies in gauge theories, *Phys. Rept.* **209**, pp. 341–378.

Shore, G. M. (1998). $U_A(1)$ problems and gluon topology: Anomalous symmetry in QCD, in D. Graudenz (ed.), *Zuoz Summer School on Hidden Symmetries and Higgs Phenomena* (PSI report PSI-PR-98-02), pp. 201–223, arXiv:hep-ph/9812354.

Shore, G. M. (2005). Strong equivalence, Lorentz and CPT violation, anti-hydrogen spectroscopy and gamma-ray burst polarimetry, *Nucl. Phys. B* **717**, pp. 86–118.

Shore, G. M. (2017). *The c and a-theorems and the Local Renormalisation Group,* Springer Briefs in Physics (Springer, Cham).

Sikivie, P. and Yang, Q. (2009). Bose-Einstein condensation of dark matter axions, *Phys. Rev. Lett.* **103**, p. 111301.

Sirunyan, A. M. *et al.* (2019). Combined measurements of Higgs boson couplings in proton–proton collisions at $\sqrt{s} = 13\,\mathrm{TeV}$, *Eur. Phys. J. C* **79**, 5, p. 421.

Sirunyan, A. M. *et al.* (2021). Evidence for Higgs boson decay to a pair of muons, *JHEP* **01**, p. 148.

Skyrme, T. H. R. (1962). A unified field theory of mesons and baryons, *Nucl. Phys.* **31**, pp. 556–569.

Slade, E. (2019). Towards global fits in EFT's and new physics implications, *PoS* **LHCP2019**, p. 150.

Smoot, G. F. *et al.* (1992). Structure in the COBE differential microwave radiometer first year maps, *Astrophys. J. Lett.* **396**, pp. L1–L5.

Solà, J. (2013). Cosmological constant and vacuum energy: Old and new ideas, *J. Phys. Conf. Ser.* **453**, p. 012015.

Solà Peracaula, J. (2022). The cosmological constant problem and running vacuum in the expanding universe, *Phil. Trans. Roy. Soc. A* **380**, p. 20210182.

Solà Peracaula, J., Gómez-Valent, A., de Cruz Pérez, J., and Moreno-Pulido, C. (2020). Brans–Dicke cosmology with a Λ-term: A possible solution to ΛCDM tensions, *Class. Quant. Grav.* **37**, 24, p. 245003.

Sozzi, M. S. (2008). *Discrete Symmetries and CP Violation: From Experiment to Theory* (Oxford University Press).

Spałek, J. *et al.* (2022). Superconductivity in high-T_c and related strongly correlated systems from variational perspective: Beyond mean field theory, *Phys. Rept.* **959**, pp. 1–117.

Sponar, S. *et al.* (2021). Tests of fundamental quantum mechanics and dark interactions with low-energy neutrons, *Nature Rev. Phys.* **3**, 5, pp. 309–327.

Straumann, N. (2002). The history of the cosmological constant problem, in *18th IAP Colloquium on the Nature of Dark Energy: Observational and Theoretical Results on the Accelerating Universe*, arXiv:gr-qc/0208027.

Straumann, N. (2004). Cosmological phase transitions, in *3rd PSI Summer School on Condensed Matter Research*, arXiv:astro-ph/0409042.

Straumann, N. (2007). Dark energy, *Lect. Notes Phys.* **721**, pp. 327–397.

Straumann, N. (2013). *General Relativity*, Graduate Texts in Physics (Springer, Dordrecht).

't Hooft, G. (1971a). Renormalizable lagrangians for massive Yang-Mills fields, *Nucl. Phys. B* **35**, pp. 167–188.

't Hooft, G. (1971b). Renormalization of massless Yang-Mills fields, *Nucl. Phys. B* **33**, pp. 173–199.

't Hooft, G. (1976a). Computation of the quantum effects due to a four-dimensional pseudoparticle, *Phys. Rev. D* **14**, pp. 3432–3450, [Erratum: *Phys. Rev. D* **18**, 2199 (1978)].

't Hooft, G. (1976b). Symmetry breaking through Bell-Jackiw anomalies, *Phys. Rev. Lett.* **37**, pp. 8–11.

't Hooft, G. (1980a). Which topological features of a gauge theory can be responsible for permanent confinement? *NATO Sci. Ser. B* **59**, pp. 117–133.

't Hooft, G. (1980b). Why do we need local gauge invariance in theories with vector particles? An introduction, *NATO Sci. Ser. B* **59**, pp. 101–115.

't Hooft, G. (2007). Emergent quantum mechanics and emergent symmetries, *AIP Conf. Proc.* **957**, 1, pp. 154–163.

't Hooft, G. and Veltman, M. J. G. (1972). Regularization and renormalization of gauge fields, *Nucl. Phys. B* **44**, pp. 189–213.

Takenaka, A. *et al.* (2020). Search for proton decay via $p \to e^+ \pi^0$ and $p \to \mu^+ \pi^0$ with an enlarged fiducial volume in Super-Kamiokande I-IV, *Phys. Rev. D* **102**, 11, p. 112011.

Taylor, J. C. (1979). *Gauge Theories of Weak Interactions*, Cambridge Monographs on Mathematical Physics (Cambridge University Press).

Terazawa, H., Akama, K., and Chikashige, Y. (1977). Unified model of the Nambu-Jona-Lasinio type for all elementary particle forces, *Phys. Rev. D* **15**, pp. 480–487.

Thorne, K. S. (2018). Nobel lecture: LIGO and gravitational waves III, *Rev. Mod. Phys.* **90**, 4, p. 040503.

Tong, D. (2016). Lectures on the Quantum Hall Effect, arXiv:1606.06687 [hep-th].

Touboul, P. *et al.* (2022). MICROSCOPE mission: Final results of the test of the equivalence principle, *Phys. Rev. Lett.* **129**, 12, p. 121102.

Trodden, M. (1999). Electroweak baryogenesis, *Rev. Mod. Phys.* **71**, pp. 1463–1500.

Tumasyan, A. *et al.* (2022). A portrait of the Higgs boson by the CMS experiment ten years after the discovery. *Nature* **607**, 7917, pp. 60–68, [Erratum: *Nature* **623**, (2023) E4].

Tumasyan, A. *et al.* (2023). Measurement of the top quark mass using a profile likelihood approach with the lepton + jets final states in proton–proton collisions at $\sqrt{s} = 13\,\text{TeV}$, *Eur. Phys. J. C* **83**, 10, p. 963.

Turok, N. (2014). Tom Kibble and the early universe as the ultimate high energy experiment, *Int. J. Mod. Phys. A* **29**, 06, p. 1430015.

Ubachs, W. (2018). Search for varying constants of nature from astronomical observation of molecules, *Space Sci. Rev.* **214**, 1, p. 3.

Uzan, J.-P. (2011). Varying constants, gravitation and cosmology, *Living Rev. Rel.* **14**, p. 2.

Van Dyck, R. S., Schwinberg, P. B., and Dehmelt, H. G. (1987). New high precision comparison of electron and positron G factors, *Phys. Rev. Lett.* **59**, pp. 26–29.

Vandersickel, N. and Zwanziger, D. (2012). The Gribov problem and QCD dynamics, *Phys. Rept.* **520**, pp. 175–251.

Veltman, M. J. G. (1968). Perturbation theory of massive Yang-Mills fields, *Nucl. Phys. B* **7**, pp. 637–650.

Veltman, M. J. G. (1981). The infrared — ultraviolet connection, *Acta Phys. Polon. B* **12**, pp. 437–457.

Veltman, M. J. G. (1997). Reflections on the Higgs system, CERN Academic Training Lecture, Yellow Report CERN-97-05.

Veneziano, G. (1979). U(1) without instantons, *Nucl. Phys. B* **159**, pp. 213–224.

Verde, L., Treu, T., and Riess, A. G. (2019). Tensions between the early and the late universe, *Nature Astron.* **3**, pp. 891–895.

Visser, M. (2019). The Pauli sum rules imply BSM physics, *Phys. Lett. B* **791**, pp. 43–47.

Volkov, S. (2019). Calculating the five-loop QED contribution to the electron anomalous magnetic moment: Graphs without lepton loops, *Phys. Rev. D* **100**, 9, p. 096004.

Vollhardt, D. and Wolfle, P. (2000). Superfluid helium 3: Link between condensed matter physics and particle physics, *Acta Phys. Polon. B* **31**, pp. 2837–2856.

Volonteri, M., Habouzit, M., and Colpi, M. (2021). The origins of massive black holes, *Nature Rev. Phys.* **3**, 11, pp. 732–743.

Volovik, G. E. (2003). *The Universe in a helium droplet*, *International Series of Monographs on Physics*, Vol. 117 (Oxford University Press).

Volovik, G. E. (2005). Cosmological constant and vacuum energy, *Annalen Phys.* **14**, pp. 165–176.

Volovik, G. E. (2006). Vacuum energy: Myths and reality, *Int. J. Mod. Phys. D* **15**, pp. 1987–2010.

Volovik, G. E. (2007). Quantum phase transitions from topology in momentum space, *Lect. Notes Phys.* **718**, pp. 31–73.

Volovik, G. E. (2008). Emergent physics: Fermi point scenario, *Phil. Trans. Roy. Soc. A* **366**, pp. 2935–2951.

Volovik, G. E. (2013). The superfluid universe, in K. H. Bennemann and J. B. Ketterson (eds.), *Novel Superfluids, Vol. 1, Int. Ser. Monogr. Phys.*, Vol. 156 (Oxford Univ. Press)), pp. 570–618.

Volovik, G. E. (2023). Gravity through the prism of condensed matter physics, *Pisma Zh. Eksp. Teor. Fiz.* **118**, 7, pp. 546–547.

von Klitzing, K., Dorda, G., and Pepper, M. (1980). New method for high accuracy determination of the fine structure constant based on quantized Hall resistance, *Phys. Rev. Lett.* **45**, pp. 494–497.

Wagner, T. A. *et al.* (2012). Torsion-balance tests of the weak equivalence principle, *Class. Quant. Grav.* **29**, p. 184002.

Wechsler, R. H. and Tinker, J. L. (2018). The connection between galaxies and their dark matter halos, *Ann. Rev. Astron. Astrophys.* **56**, pp. 435–487.

Weinberg, D. H. and White, M. (2024). Dark energy, in S. Navas *et al.* (eds.), *The Review of Particle Physics (2023)*, [Particle Data Group] (Chapter 28), Phys. Rev. D **110** (2024), 030001.

Weinberg, S. (1967). A model of leptons, *Phys. Rev. Lett.* **19**, pp. 1264–1266.

Weinberg, S. (1972). *Gravitation and Cosmology: Principles and Applications of the General Theory of Relativity* (John Wiley and Sons, New York).

Weinberg, S. (1976). Critical phenomena for field theorists, in A. Zichichi (ed.), *14th International School of Subnuclear Physics: Understanding the Fundamental Constitutents of Matter* (Springer, Boston, MA).

Weinberg, S. (1978). A new light boson? *Phys. Rev. Lett.* **40**, pp. 223–226.

Weinberg, S. (1979). Baryon and lepton nonconserving processes, *Phys. Rev. Lett.* **43**, pp. 1566–1570.

Weinberg, S. (1987). Anthropic bound on the cosmological constant, *Phys. Rev. Lett.* **59**, pp. 2607–2610.

Weinberg, S. (1989). The cosmological constant problem, *Rev. Mod. Phys.* **61**, pp. 1–23.

Weinberg, S. (1995). *The Quantum Theory of Fields. Vol. 1: Foundations* (Cambridge University Press).

Weinberg, S. (1996). *The Quantum Theory of Fields. Vol. 2: Modern Applications* (Cambridge University Press).

Weinberg, S. (2008). *Cosmology* (Oxford University Press).

Weinberg, S. (2018). Essay: Half a century of the Standard Model, *Phys. Rev. Lett.* **121**, 22, p. 220001.

Weinberg, S. and Witten, E. (1980). Limits on massless particles, *Phys. Lett. B* **96**, pp. 59–62.

Weiss, R. (2018). Nobel lecture: LIGO and the discovery of gravitational waves I, *Rev. Mod. Phys.* **90**, 4, p. 040501.

Wells, J. D. (2009). Lectures on Higgs boson physics in the Standard Model and beyond, in *38th British Universities Summer School in Theoretical Elementary Particle Physics*, arXiv:0909.4541 [hep-ph].

Wen, X.-G. (2002). Origin of light, *Phys. Rev. Lett.* **88**, p. 011602.

Wen, X.-G. (2004). *Quantum Field Theory of Many-Body Systems* (Oxford University Press).

Wen, X.-G. (2013). Topological order: From long-range entangled quantum matter to an unification of light and electrons, *ISRN Cond. Matt. Phys.* **2013**, p. 198710.

Wess, J. and Zumino, B. (1974). A lagrangian model invariant under supergauge transformations, *Phys. Lett. B* **49**, pp. 52–54.

Wetterich, C. (1988). Cosmology and the fate of dilatation symmetry, *Nucl. Phys. B* **302**, pp. 668–696.

Wetterich, C. (1995). The Cosmon model for an asymptotically vanishing time dependent cosmological 'constant', *Astron. Astrophys.* **301**, pp. 321–328.

Wetterich, C. (2007). Growing neutrinos and cosmological selection, *Phys. Lett. B* **655**, pp. 201–208.

Wetterich, C. (2010a). Fermions from classical statistics, *Annals Phys.* **325**, pp. 2750–2786.

Wetterich, C. (2010b). Quantum mechanics from classical statistics, *Annals Phys.* **325**, pp. 852–898.

Wetterich, C. (2017). Gauge symmetry from decoupling, *Nucl. Phys. B* **915**, pp. 135–167.

Wetterich, C. (2021). Quantum fermions from classical bits, *Phil. Trans. Roy. Soc. A.* **380**, 2216, p. 20210066.

Wetterich, C. (2022). Fermionic quantum field theories as probabilistic cellular automata, *Phys. Rev. D* **105**, 7, p. 074502.

Wheeler, J. A. and Ford, K. (1998). *Geons, Black Holes, and Quantum Foam: A Life in Physics* (W. W. Norton & Company, N.Y.).

Wilczek, F. (1978). Problem of strong P and T invariance in the presence of instantons, *Phys. Rev. Lett.* **40**, pp. 279–282.

Wilczek, F. and Zee, A. (1979). Operator analysis of nucleon decay, *Phys. Rev. Lett.* **43**, pp. 1571–1573.

Wilczek, F. and Zee, A. (1984). Appearance of gauge structure in simple dynamical systems, *Phys. Rev. Lett.* **52**, pp. 2111–2114.

Will, C. M. (2014). The confrontation between general relativity and experiment, *Living Rev. Rel.* **17**, p. 4.

Wilson, K. G. and Kogut, J. B. (1974). The renormalization group and the epsilon expansion, *Phys. Rept.* **12**, pp. 75–199.

Witten, E. (1979). Current algebra theorems for the U(1) goldstone boson, *Nucl. Phys. B* **156**, pp. 269–283.

Witten, E. (1980). Large N chiral dynamics, *Annals Phys.* **128**, pp. 363–375.

Witten, E. (2018). Symmetry and emergence, *Nature Phys.* **14**, 2, pp. 116–119.

Yanagida, T. (1979). Horizontal gauge symmetry and masses of neutrinos, *Conf. Proc. C* **7902131**, pp. 95–99.

Ye, J. and Zoller, P. (2024). Essay: Quantum sensing with atomic, molecular, and optical platforms for fundamental physics, *Phys. Rev. Lett.* **132**, 19, p. 190001.

Yunes, N., Miller, M. C., and Yagi, K. (2022). Gravitational-wave and X-ray probes of the neutron star equation of state, *Nature Rev. Phys.* **4**, 4, pp. 237–246.

Zaanen, J. and Beekman, A. J. (2012). The emergence of gauge invariance: The Stay-at-home gauge versus local-global duality, *Annals Phys.* **327**, pp. 1146–1161.

Zeldovich, Y. B. (1967). Cosmological constant and elementary particles, *JETP Lett.* **6**, pp. 316–317.

Zeng, B., Chen, X., Zhou, D.-L., and Wen, X.-G. (2019). *Quantum Information Meets Quantum Matter — From Quantum Entanglement to Topological Phase in Many-Body Systems* (Springer).

Zinn-Justin, J. (1989). *Quantum field theory and critical phenomena, International Series of Monographs on Physics*, Vol. 77 (Oxford University Press).

Zinn-Justin, J. (2007). *Phase Transitions and Renormalization Group* (Oxford University Press).

Zohar, E., Cirac, J. I., and Reznik, B. (2016). Quantum simulations of lattice gauge theories using ultracold atoms in optical lattices, *Rept. Prog. Phys.* **79**, 1, p. 014401.

Zwanziger, D. (1989). Local and renormalizable action from the Gribov horizon, *Nucl. Phys. B* **323**, pp. 513–544.